AF576964

COST-EFFECTIVE MAINTENANCE MANAGEMENT

This book is dedicated to my beloved wife, Linda, who taught me all the important things about life, none of which are taught in school.

COST-EFFECTIVE MAINTENANCE MANAGEMENT

Productivity Improvement and Downtime Reduction

by

Frank Herbaty

Frank Herbaty & Associates
Maintenance Management Consultants
Bolingbrook, Illinois

np

NOYES PUBLICATIONS
Park Ridge, New Jersey, USA

Library of Congress Catalog Card Number: 83-13072
ISBN: 0-8155-0953-7
Printed in the United States

Published in the United States of America by
Noyes Publications
Mill Road, Park Ridge, New Jersey 07656

10 9 8 7 6 5 4 3 2 1

Library of Congress Cataloging in Publication Data

Herbaty, Frank.
Cost-effective maintenance management.

Bibliography: p.
Includes index.
1. Plant maintenance. 2. Industrial equipment--
Maintenance and repair. I. Title.
TS192.H47 1983 658.2'02 83-13072
ISBN 0-8155-0953-7

Preface

My love affair with maintenance started when I graduated from college in 1944. I joined the Merchant Marine and worked in the engine room. The chief engineer soon found out that I didn't know the difference between a monkey wrench and a Stillson wrench. This, coupled with the fact that I was a graduate mechanical engineer, earned me his undying contempt and frequent beatings with his fists when I did or said anything to displease him. I might add that he was about 6′ 6″, 260 pounds and built like Hercules! In those days, before 100% unionization, the captain and chief engineer literally had a life and death control over the crew. If you made them mad, watch out!

I learned fast, and rose to the rank of chief engineer in two and one half years, something impossible to do during non-war years, when it might normally take at least ten years. Back in those days, each engineer had to know how to operate and maintain everything, including mechanical, electrical, electronic, structural, radio, and all other types of equipment aboard ship. After a year, my pulse quickened every time I walked down into the engine room. there were the characteristic smells, sounds and vibrations: steam, oil, heat, water, pumps, motors, boilers, evaporators. I got to the point where I could tell by listening and by smelling if everything was O.K. by the instantaneous integration of the hundreds of different sounds and smells. I was hooked, I was in love with maintenance.

Since those early days my maintenance knowledge has been constantly growing. Everytime I think that I know it all, I learn more. Everytime I become certain of how to do something, I learn a better way. Will it ever end? I don't think so. How long does it take to design, implement and analyze the worth of a preventive maintenance program? Ten years? And how does *your* system compare with someone else's system? There really isn't enough time in life to find out. Besides, Plant Engineers don't stay long enough with a company to find out. The person who has the best chance of finding the best way is the maintenance management consultant. He has the opportunity of seeing and evaluating many different systems and is, therefore, in the best position to pick a winner.

It is important that you know something about my background so that you will know where I am coming from and that you will be convinced that I have the credentials to tell you that *I have found the answer to optimization of the maintenance management function.* That is what this book is all about. This is my background:

I have been a maintenance management consultant and have held many other jobs, all involved with maintenance. The variety of jobs has given me a view of maintenance from every possible angle and has helped me in my understanding of what is, and is not important. My experience covers consulting engineering (10 years), marine engineering (5 years), plant engineering, maintenance and production (16 years), management consulting (4 years) and research (5 years), total experience 40 years.

In these categories I have held many high level positions including Vice President of Pope, Evans and Robbins, a well known consulting engineering firm, Plant Engineering Manager for Rohr Corporation, Director of Engineering, Operations and Maintenance at Kennedy Space Center, Director of Research for President Carter's Reorganization Project, Senior Program Manager at the Department of Energy for three large research programs and was cited by a leading trade journal as being one of the top five maintenance consultants in the country.

In giving lectures and seminars throughout the world, I have found the telling of anecdotes from my past experience to be very effective in illustrating the point being discussed. For this reason, I have made liberal use of this technique as case studies in the text. I have also decided to use the informal approach in writing and will therefore, write as if I am speaking to you in my living room. I feel confident that the use of the optimization principles expounded in this text will bring dramatic improvement to any maintenance organization.

One more point. I don't intend to philosophize about alternate ways of doing things. I've paid my dues and made my choice and have selected the best system for each function to the best of my knowledge. As a result, only one system is described for each function. Although the system described may not exactly fit every plant, it can easily be modified to suit any plant. I have had no problem doing so on any of my assignments. My use of one system has its advantages since I go into sufficient detail for the reader to implement the system.

This handbook is divided into five parts. A brief explanation of the content of each of these parts follows:

Part 1

This part covers the approach to maintenance management optimization. It defines what maintenance is and how its cost is measured and states that effort must be applied in proportion to the impact the effort has on:

- Company goals and objectives (or mission);
- Profit, cost effectiveness and life cycle cost;
- Safety;
- Environmental impacts; and,
- Product quality (or mission quality).

Parts 2 and 3

Parts 2 and 3 discuss downtime and productivity control. These are two out of possibly twenty subjects that may be discussed under the subject of maintenance management. I have found, beyond a shadow of a doubt, that the importance of these two subjects totally dwarfs the impact the other subjects have on the maintenance management function.

Part 4

Recognizing that much has been written on maintenance support functions, such as material, engineering, purchasing, accounting, computers, personnel and the union, etc., I confine my remarks in Part 4 to things that I have learned that are unique or helpful in managing these support functions.

There are a few books in existence on maintenance management. They include all functions that are legitimately a part of the subject. This is the first book however, that emphasizes optimization and differentiates between the trivial and the two most important functions of a maintenance manager by presenting a practical system for optimizing maintenance and concentrating corrective effort where it will reap the highest rewards.

Bolingbrook, Illinois Frank Herbaty
June 1983

Contents and Subject Index

Part 1

MAINTENANCE OPTIMIZATION APPROACH

1

General

Maintenance is performed to help achieve the goals and objectives of a company or organization. In the case of a manufacturing organization, the objective is to produce and sell a product at a profit. In many organizations, this objective is simply to perform a mission in a cost-effective manner. For example, a mission of the strategic air command is to keep the required number of bombers in the air at all times in a cost-effective manner. The maintenance optimization principles described in this text apply equally as well to any type of organization, be it product or mission oriented, (the latter being the case in most government organizations).

Optimization of maintenance is the art of performing maintenance, within a set of constraints, in a manner that best achieves the goal of the company. One set of constraints might be to state the sum of maintenance labor and material, *plus* production loss costs caused by inadequate maintenance, must be optimized to the minimum value. The trouble with this simplistic view is that other things must also be considered, for example, safety and environmental impact. If all factors are considered in the art of optimization, it becomes very complex.

The optimization approach about to be described has the following characteristics:

- There are many optimization systems, not one. Each approaches the problem in a slightly different manner.
- Most all of the optimization systems are cyclical in nature and are designed to lessen the problem each time a cycle is complete until finally, there is no more problem.
- The optimization systems recognize that the primary task of the maintenance department is to perform maintenance work and therefore the manhours for analysis are hard to come by and must not be squandered on non-productive effort. For this reason a very important part of the optimization system is to establish methods for determining criticality of both effort and facilities and equipment.
- Finally, downtime control (Part 2) and productivity control (Part 3) are of such importance in optimizing maintenance, that the entire emphasis in this text is placed on these two subjects. This is done because the other maintenance subjects either have trivial impact or they are normally not a problem.

With these thoughts in mind several of the most important general optimization systems are described, keeping in mind that downtime and productivity control systems (Parts 2 and 3) are optimization systems also. It should also be noted that one can enter into each cycle at any point, since the cycle repeats itself many times. This text, however, is written from the viewpoint of an existing organization within existing goals and objectives, rather than from the viewpoint of a plant starting up for the first time.

2

Audit Priorities

Every manager should perform periodic audits to ensure that existing goals, objectives, policies and procedures are being properly executed. It is recommended that the following auditing priority system be followed to ensure that analytical effort be applied in the proper sequence and to insure that this effort is not squandered on the wrong function. (For example, there would be no point in auditing job methods when a crew is spending 30% on travel time, 30% material acquisition time, 10% obtaining tools, 20% idle and only 10% of the time working. In this case, audits should be conducted on travel and material acquisition times first).

PRIORITY 1–CONTROL COMPETENCE OF MAINTENANCE LABOR

The first priority is to insure the availability of technically competent mechanics who can repair anything that breaks down. This is the most obvious to the owner of a small plant who is about to hire his first maintenance man. There is no quesiton in *his* mind that this is the first priority.

PRIORITY 2–CONTROL OVER THE "LOGIC" OF WORK ASSIGNMENTS

This item can best be illustrated by giving a few examples of illogical work assignments that should be caught and stopped by the audit, i.e.,

Construction Work: It is illogical to have maintenance men build a desk for $300, when a superior desk can be purchased for $200.

Janitorial Work: It is illogical to have an electrician clean offices when the work should be performed by a lower paid janitor.

Design Changes: It is illogical to make design changes or modifications by unqualified operation or maintenance personnel. Checks and balances must be instituted to insure that authorized and qualified personnel work on appropriate work.

Maintenance Work: It is illogical to perform maintenance work that can be eliminated by correcting a design error (replacing a fuse that is constantly blowing with one of a correct size).

It is illogical to perform maintenance work that can be eliminated by correcting an operational practice that causes damage to facilities (craneman damaging structural elements by collision).

It is illogical to perform repetitive maintenance tasks that can be eliminated by a revision to preventive maintenance procedures. It is important to realize that some repetitive maintenance procedures can become so ingrained over the course of many years that the illogicalness of the procedure is overlooked.

PRIORITY 3–CONTROL OF PRODUCTION LOSS CAUSED BY EQUIPMENT DOWNTIME

The cost of production loss caused by equipment downtime must be minimized by the optimization techniques described in Part 2.

PRIORITY 4–CONTROL OVER LABOR PRODUCTIVITY

Then, systems must be designed to obtain high labor productivity as described in Part 3.

PRIORITY 5–MANAGEMENT INFORMATION AND CONTROL SYSTEMS

Finally, management information and control systems must be installed to "fine tune" the maintenance function.

PROBLEMS IN AN EXISTING ORGANIZATION SHOULD BE CORRECTED IN THE ABOVE ORDER OF PRIORITY!

3

"Management" as an Optimization System

The functions performed by management, constitute an optimization system of the greatest importance. The functions are cyclical in nature and are repeated over and over again, never ending, each time, improving the total performance of the organization. The functions of management are:

(1) Establish goals and objectives, policies and procedures.

(2) Establish permissible variance from these guidelines.

(3) Measure performance to established guidelines.

(4) Compare performance measurement feedback information to guidelines.

(5) Isolate and identify deviations beyond tolerances.

(6) Determine basic cause for deviations.

(7) Determine corrective action.

(8) Plan method of implementing corrective action.

(9) Schedule plan for implementing corrective action.

(10) Implement corrective action.

(11) Follow-up to ensure completion of corrective action and to prevent overswing.

These functions are shown graphically in Figure 1-1.

Stated more briefly, management must establish goals and objectives then design a performance measurement system to measure performance to these goals and objectives. The goals and objectives and the policies and procedures written to implement them may be a long and complex list of guidelines. Just writing the guidelines will not insure that they are implemented. They may not be implemented for one or all of the following or other reasons.

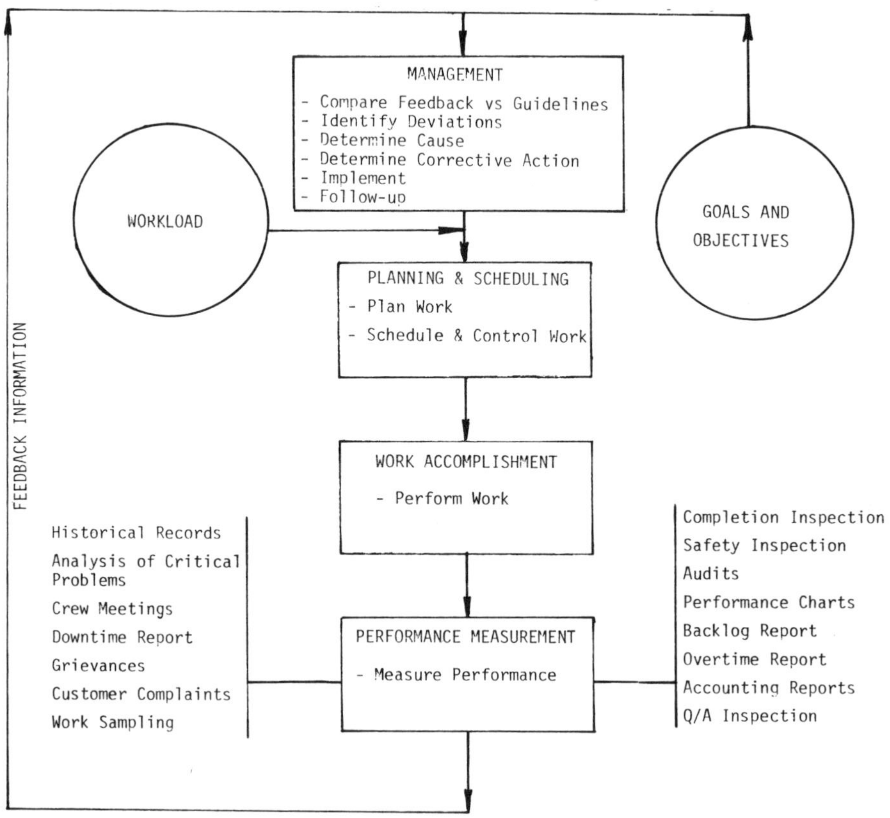

Figure 1-1: Functions of management.

- Employees don't understand the guideline
- Employees don't like the guideline
- So many procedures have been written that the employee doesn't have time to carry out all of them
- Employee likes his way better
- Etc.

Managers must perform periodic audits of all guidelines and study feedback reports, otherwise employees will be doing what they want to do, which may be nothing. Another good time to audit procedures is during a major crisis, breakdown or accident, since violation of established guidelines is frequently the cause. A few case studies by the author will illustrate the point.

CASE STUDY 1A: The Frozen Safety Valve

As was my practice at company x, I periodically performed random audits of procedures. I was auditing the functions of a material expeditor. I closed my eyes and reached into the drawer and took out a purchase requisition at random from the expeditor's desk. It was marked "Urgent Safety Item" and was three weeks old. It was an order for a new safety valve for the cafeteria boiler. Investigation revealed that a mechanic, while performing a check by lifting the safety valve levers, had found it frozen. He immediately reported this to his foreman; however, nothing had been done about the problem in three weeks. I immediately ordered the boiler to be shutdown and the valve overhauled; a task which was completed in four hours. At least four company procedures were violated. Both of these employees were severely disciplined for this serious infraction of company rules. I had previously led an investigation of a boiler explosion in a cafeteria which killed dozens of people because the engineer forgot to remove boiler safety valve gags.

CASE STUDY 1B: The Burning of a Welder

A welder had just completed burning out a section of floor plate inside an electrical cubicle. When he was finished, he lifted up the cut-out section and let it fall against high voltage bus-bars.

This caused an explosion and severe burning of the welder by flying pieces of molten copper. Investigation of this incident revealed that the following company procedures had been violated:

- Unauthorized entry into a cubicle sealed by Q/A
- Maintenace performed during a launch countdown without the Director's written approval
- The foreman had failed to brief his crew on the safety hazards of the job and failed to initial the box on the work order certifying that he had done this
- The welder was not supposed to be in the cubicle without being with a journeyman electrician
- The power was not shut off
- Written approval for adding a new line to the cubicle had not been obtained
- The job was started by the 8 to 4 shift foreman who failed to brief the 4 to 12 shift foreman on the job status.

In summary, it must be emphasized that the management control cycle is a continuous, never ending optimization process, which requires constant surveillance on the part of the manager. In order for it to work, every department must have established goals, objectives, policies, procedures and a matching and well thought out performance measurement system.

4

Maintenance Goal

The goal of any well run maintenance organization is to have the lowest cost of the sum of two quantities, i.e:

- Maintenance labor and material; and
- Production loss cost resulting from an inadequate maintenance program.

The achieving of the lowest cost is an optimization technique and is shown graphically in Figure 1-2. Notice that when maintenance cost is zero (no maintenance) the production loss cost is naturally at its highest. As maintenance effort is intelligently increased, the production loss gradually decreases until the lowest combined cost is achieved. This is the maintenance goal. Maintenance effort applied beyond this point, *increases* cost.

This phenomenon is true of any service organization; for example, purchasing and stores. Its costs are also the sum of its labor and material costs, plus, the losses caused by inadequate service to its client.

Unfortunately, this fact is not often recognized or accepted in industry as is illustrated by the following case studies.

CASE STUDY 1C: Who Should Be Charged?

While doing a fleet management study, I noticed that the downtime on all construction vehicles used to put in new elec-

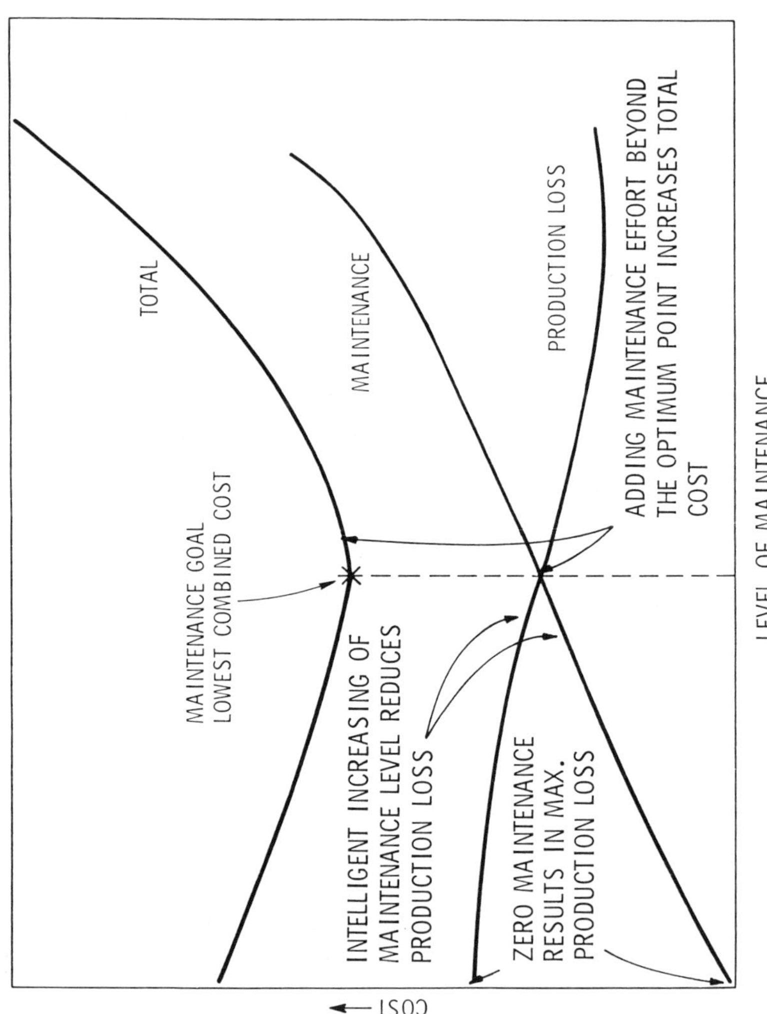

Figure 1-2: Maintenance goal.

tric and gas services and transmission lines was excessive. When a crew of thirteen workers was dispatched to a job, a critical vehicle frequently broke down (such as a bucket truck or pole hole driller). The crew would usually wait for someone to pick them up and tow the vehicle back to the shop. This normally wasted the entire day for the crew while waiting for their vehicle to be fixed. The lost time was charged to the construction work order.

I argued that this method of charging time did not put pressure where corrective effort was required; on the fleet manager. I won the battle and thereafter, lost time caused by vehicle breakdowns, was charged to the fleet manager, who was then raked over the coals at weekly meetings to reduce this lost time cost. He got the message and an improving trend was soon noticed.

CASE STUDY 1D: The Efficient Purchasing Manager?

At a certain plant, I was doing a study of the engineering, maintenance, purchasing and stores departments. A quick survey of the maintenance department personnel quickly surfaced material acquisition as their biggest problem. This was not surprising, since the mine and plant were in an isolated, remote area. Excessive stock outages quickly led me to the purchasing department.

The purchasing manager was indignant about being audited and was proud of the fact that he had just made a study of his department which resulted in his laying off ten out of the 20 people who worked for him. This was done at the prodding of top management who were anxious to reduce indirect costs.

I asked him if he had made any attempt to determine the impact his layoff had on the 5,000 people he served. He hadn't of course, and gruffly mentioned that this was none of his business and that his job description called for him to run an efficient *purchasing* department.

He was not sensitive to the concept shown in Figure 1-2. Needless to say, his concept of "purchasing efficiency" changed shortly, since his department's performance was the major factor inhibiting adequate maintenance performance.

Figure 1-3 shows the two control systems used to control the two costs shown in Figure 1-2, i.e:

Cost	Control System
Maintenance labor and material	Productivity
Production loss	Downtime

The intelligent application of these two control systems has a greater influence on the success of a maintenance organization than any other factor.

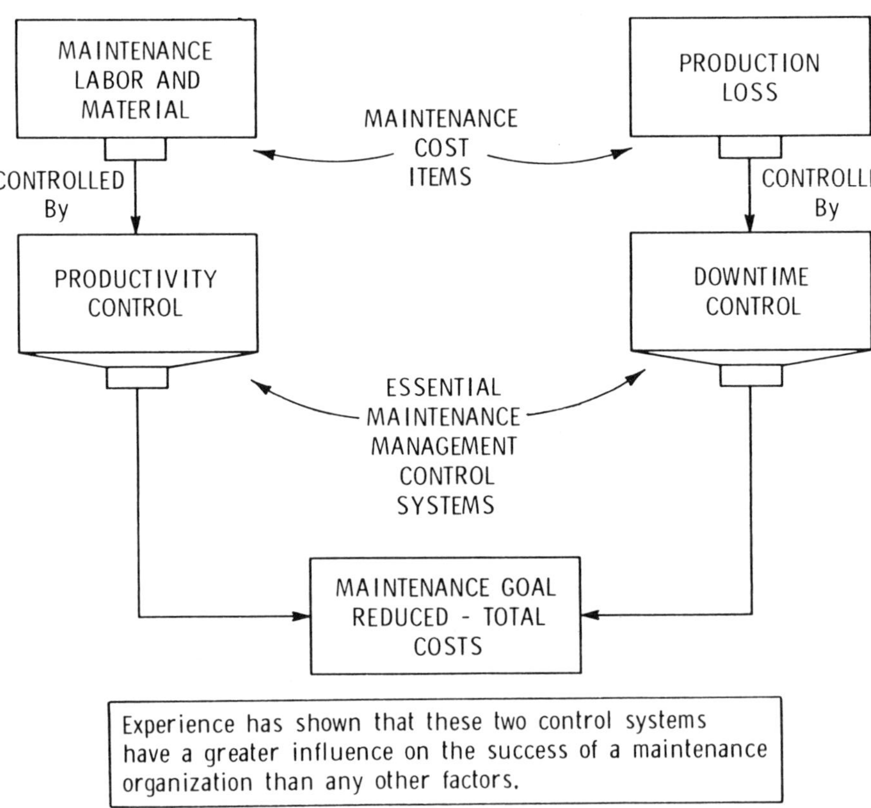

Figure 1-3: The two essential control systems.

5

Criticality Determination

GENERAL

Obviously not all equipment, materials, functions, and labor efforts are of equal importance relative to the mission, or goals and objectives of a particular organization. This fact *should* lead all of us to the following conclusions:

- It is of the utmost importance that everyone in the company understand fully the mission, or goals and objectives of the company
- That equipment, materials, functions and labor effort should be categorized by criticality to the company's mission, goals and objectives, so that emphasis can be placed on those items having the highest criticality.

Although the author has been in the business for many years, he has yet to find a company that does an effective job at the two tasks listed above. Even worse, most companies don't even make the attempt. Presented here are several methods of categorizing by criticality and examples that illustrate how the methods are used, what happens when you don't do this and company employees are not familiar with the company's objectives. (In general, the impact is a squandering of the company's valuable labor and material resources and low productivity!)

NUMBERS OF ITEMS VERSUS COST OR LABOR HOURS

In a certain plant 80% of the work orders accounted for only 10% of the costs and 20% of the work orders accounted for 90% of the costs!

This reverse relationship is true of most functions in an organization! Knowledge of this fact is of tremendous value in making criticality determinations. The relationship can be surfaced by writing a very simple computer program that sorts information already in the computer to facilitate criticality determinations. Figure 1-4 illustrates the technique. Notice that in this figure, only one half (0.5) of 1% of the number of equipment items out of 1,000 account for 77% of the annual maintenance cost. Obviously, these five equipment items should receive the lion's share of maintenance effort if cost alone is used as a criteria. The following illustrates the use of charts such as Figure 1-4.

Criticality Analysis—A Case Study

A chart identical to that shown in Figure 1-4 was made for a process plant being studied. The chart revealed the following:

- 1% of 10,000 equipment items accounted for 80% of the annual maintenance cost.
- 50% of the 10,000 items accounted for only one tenth (0.1) of 1% of the annual costs.

Further study revealed that the preventive maintenance (PM) program was applied uniformly over all 10,000 items. The following action was taken:

- 50% of the items were removed from the computer and preventive maintenance programs and put on a 100% breakdown maintenance program, except for those items requiring lubrication. This greatly simplified the computer run analysis and permitted emphasis to be given to the more important equipment.
- The remaining 50% of the items were coded by criticality representing decreasing mission criticality, and the codes entered into the

Item Number	Machine Number	Description	Annual Maintenance Cost Dollars	Annual Maintenance Cost Percent	Cumulative Cost Percent	Cumulative Number of Items Percent
1	001	-	xxx	30	30	0.1
2	008	-	xxx	29	59	0.2
3	109	-	xxx	10	69	0.3
4	110	-	xxx	5	74	0.4
5	005	-	xxx	3	77	0.5
6	854	-	xxx	1	78	0.6
7	562	-	xxx	1	79	0.7
8	687	-	xxx	0.4	79.4	0.8
9	295	-	xxx	0.3	79.7	0.9
10	346	-	xxx	0.3	80.0	1.0
↓	ETC	-	ETC	ETC	ETC	ETC
↓	ETC	-	ETC	ETC	ETC	ETC
1000	↓		↓	↓	↓	↓
TOTALS			xxx	100	100	100

Figure 1-4: Criticality determination.

computer program as part of the equipment number. A program was initiated to tailor the thoroughness of the preventive maintenance program to the criticality of the equipment.

- It was decided that 100% of the maintenance analysis effort would be put immediately on the 1% of the items accounting for 80% of the cost.
- It was decided to make a chart similar to Figure 1-4 for all 30,000 line items of stock and another chart for downtime and production loss cost versus equipment item. (Parts usage is a barometer of equipment failure).

The new charts revealed the following:

- Ten items of stock out of 30,000 accounted for 70% of the annual stock replacement cost.
- 20,000 items of stock accounted for only 1% of the annual cost.
- 200 out of 10,000 equipment items accounted for 85% of the annual downtime hours.
- Only two out of the 10,000 equipment items accounted for 100% of the production loss cost!

The following action was taken as a result of the new charts:

- A special task force was assigned to investigate the ten stock items accounting for 70% of the cost with the objective of either solving the wear or failure problem, obtaining lower prices or otherwise, reducing costs. These ten items were found to coincide with the 1% of the equipment items causing 80% of the maintenance cost.

  10,000 items were eliminated from stock and 500 items were transferred to open bench stock.
- A special system was designed by myself for investigating the startling revelation that only

two equipment items accounted for 100% of the production loss cost. (See part 2, "The principal of simultaneous downtime"). It is interesting to note that downtime does not necessarily cause a production loss. Reasons for downtime not causing a production loss may be:

— redundant equipment available
— equipment is producing into in-process storage
— stockpiles in the distribution system
— plant was not operating 24 hours per day, 7 days per week and therefore, loss could be made up.

CASE STUDY 1E: The Crusher Balls

An analysis of spare parts costs at a plant was made similar to that shown in Figure 1-4. This chart revealed that crusher balls (4″ diameter steel balls) were one of the three highest cost items and cost approximately $5,000,000 per year to replace. These balls were used in 25, 1,000 HP ore crushers. The balls were purchased from six suppliers. It was decided to run a test by dividing the crushers into six groups and then using one supplier for each of the six groups. Cost of balls per ton of throughput was measured over a period of one month. Each ball varied slightly in material composition and manufacturing technique.

The results of the test showed that one ball cost 20% less per ton of throughput. Orders for others were cancelled and the optimum ball was used exclusively thereby, saving $1,000,000 per year.

CASE STUDY 1F: What Is Your Mission?

Most government agencies have their missions described in a public law passed by Congress. Some of the larger agencies have multiple and complex missions. Although the missions are clear in the law and in the top echelons of management, it is not uncommon for the worker three or four levels down in the organization structure to completely lose sight of the agencies mission. The following example will illustrate the point:

It was observed that one very large government agency, which had a maintenance mission, had several hundred forms with accompanying procedures, which had to be filled out as part of performing the various maintenance functions. In interviewing the supervisor in charge of forms and procedures and asking him what his mission was, he stated that his mission was to generate forms and procedures as stated in his job description. This is a common fault in many job descriptions, which fail to even mention or make clear what the mission of the organization is.

It was explained to the supervisor that his function was indirect staff support to the direct mission, which was maintenance. Indirect support can only be justified by means of either improving the productivity of direct labor or by fulfilling a mandatory function such as safety. For example, when a foreman is hired to supervise fifteen maintenance people, it is done in anticipation of his increasing work productivity by say, 20%, thereby saving three workers for the price of one supervisor. In a similar manner, it was explained that it was the supervisor's job to write forms and procedures only to the extent that they increase maintenance labor productivity. It was quite obvious to anyone knowledgeable about maintenance management that the issuance of several hundred forms and procedures would actually *decrease* productivity soon after the number of forms exceed ten or twenty. In a similar manner, the safety engineer may think that safety is the company's mission, the procurement specialist thinks that it's his job to uphold the integrity of the procurement system, and the personnel man, the integrity of the personnel system, etc., etc.

Plant personnel in the case studies thought they had a good feel for the criticality of the various maintenance factors involved, but were surprised at the actual statistics when plotted as previously described. This lack of specific knowledge of plant criticality factors is more often the case than not. This approach can and must be used in *all* areas of maintenance management if effective use of limited personnel is to be made.

Suggestions for other systems to be analyzed in this manner are given in the following tabulation:

- Number of downtime occurrences versus equipment items
- Maintenance cost per ton or per square feet or per dollar value of facility versus department
- Stock parts cost versus specific part
- Dollars production loss versus equipment item
- Job type versus cost
- Analysis of repetitive failures by equipment item, failure type, cost per failure and failure frequency
- Numbers of equipment items versus criticality code
- Productivity percentages versus maintenance area
- Productivity percentages versus maintenance trade
- Productivity percentages versus foreman
- Productivity percentages versus job type
- Union grievances by type, foreman and trade.

CRITICALITY CODING

Each plant has many thousands of parts, machines, people, and problems. If trivia are not separated from critical items, the system will collapse under its own mass of data, paperwork, and instructions.

Therefore, it is very important to code equipment by criticality. The code should be added to the equipment number to facilitate sorting by the computer. Only after criticality is determined should maintenance efforts be applied. Several factors should be considered in determining criticality:

Dollar Value: A $500,000 piece of equipment is more important than a $100 piece of equipment and should receive greater maintenance care.

Function: Equipment vital to production should receive greater maintenance attention, particularly if downtime will result directly in lost production.

Constant Maintenance Problems: Equipment that frequently breaks down should receive greater attention.

Other factors that affect criticality will depend on problems relating to a specific plant's needs, and the criticality coding system should be tailored to the particular facility. Examples of two simple coding systems follow:

Code A—Critical production equipment for which there is no backup or substitute.

Code B—Critical production equipment with back-up or in-process storage.

Code C—Noncritical equipment.

or

Code 1—Critical equipment–downtime may result in loss of sales.

Code 2—Semicritical equipment—downtime may result in lost production that can usually be recovered.

Code 3—Noncritical equipment—downtime seldom affects production.

Once the coding system is designed, maintenance efforts should be apportioned according to criticality. Such effort will be in the form of:

- Breakdown maintenance labor and materials

- Preventive maintenance labor and materials
- Design of maintenance optimization systems
- Maintenance engineering analysis.

If breakdowns on ten of 10,000 equipment items are causing 99% of the production loss, a major portion of the maintenance analysis effort should be spent on these 10 items. Also, the Preventive Maintenance program on these items should be strengthened accordingly.

6

Type of Organization

The percentage that maintenance cost is of the total operating budget varies over a very wide range. For example, a service organization may rent office space where the rental cost includes maintenance. This organization would have no maintenance people on their payroll.

Many manufacturing organizations have maintenance costs that vary between 5–10%.

On the other hand, most process plants have maintenance costs that vary between 50–95% and the maintenance percentage of these and other types of plants is increasing due to automation and robotization.

The maintenance manager should be aware of the maintenance percentage for several reasons:

- He must be aware of the percentage of his competitors. If his cost is higher, the lower cost company may run his company out of business.
- He must be aware of the availability percentage of critical equipment, lest competitors run him out of business with higher percentages.
- If the maintenance percentage is running around 1%, none of top management will have much interest in sophisticated maintenance

systems. If on the other hand, the maintenance percentage is about 95%, top management will be scrutinizing his activity, perhaps on a daily basis, and if he does a good job, he may eventually become general manager of the plant. The maintenance manager, with costs at 1%, may never achieve a higher position.

7

Preventive Maintenance (PM) Policy

GENERAL

The determination of which equipment items should be included in the preventive maintenance program and the degree of preventive maintenance is part of the maintenance optimization program. The methods of determining criticality previously discussed should be the basis for making this determination. Lack of an intelligent analysis of these two factors will increase maintenance costs. The large number of companies that decide to include *all* equipment when starting a preventive maintenance program for the first time is astonishing!

All members of a task force charged with designing and implementing a new preventive maintenance program should be aware of the following facts:

- All effort expended on designing a new preventive maintenance program is an expense subtracting from profit.
- Preventive maintenance tasks will *increase* maintenance costs when first initiated, until the beneficial effect of the preventive maintenance task has time to take effect.

- A preventive maintenance task may *permanently* increase costs, if the author of the task doesn't know enough about maintenance. (This point is made because some organizations assign the function to unknowledgeable or clerical personnel under the mistaken theory that all they have to do is copy the instruction from the manufacturers maintenance manuals.)
- Finally, at the start of the preventive maintenance program, people must be selected to both design and implement the program. Normally these people already have jobs and there may be pressure, in some cases, to assign the least competent personnel available. This tendency must be vigorously combated, particularly in the case of people assigned to design the system, who should be the *most competent* people in the organization.

With these points in mind, management should proceed as follows:

- Assign the most competent people to design the system.
- Start with the most critical equipment (determined from a computer printout by criticality code), and schedule, the instruction writing and implementation tasks in accordance with the manpower available, incrementally, so as not to overtax the available manpower.
- The implementation rate should be designed to give time for the preventive maintenance instruction to reap its benefit in reduced labor, material and production loss costs. Ideally, the preventive maintenance task labor added should equal the savings from previously installed preventive maintenance. In this manner, the only additional effort required is for the first small group of preventive maintenance tasks implemented. In other

> words, preventive maintenance labor is substituted for breakdown maintenance labor on an incremental and equal basis. By the time the entire preventive maintenance is implemented, a net savings in total maintenance cost should be realized.

Optimization of the preventive maintenance instruction is covered in detail in chapter 12. The following paragraphs give examples of the different choices available when making the decision on the degree of thoroughness varying from zero preventive maintenance to a preventive maintenance program that is designed for zero failures.

100% BREAKDOWN MAINTENANCE

This equipment is handled by the trouble call system. Nothing is done unless someone calls the trouble desk to announce equipment failure. An example might be an exhaust fan in a conference room with lifetime lubricated bearings.

BREAKDOWN MAINTENANCE WITH MINOR LUBRICATION ONLY

This case is the same as the preceding paragraph, except that the fan bearings require periodic lubrication. The lubrication would be handled as part of a preventive maintenance program. A car owner who plans to buy a new car every 1—2 years could use this system performing no preventive maintenance other than checking the engine oil and adding oil when necessary.

MINOR PREVENTIVE MAINTENANCE

Many maintenance experts advocate use of this system exclusively which consists of:

- Necessary lubrication
- Preventive maintenance inspection, adjustments and repair limited to a maximum time

per machine of say, ten minutes. Anything over this time will be reported to the foreman and be covered by a special work order.

The number of trouble calls and reports of problems during the inspection serve as an indicator of the adequacy of preventive maintenance instruction.

MAJOR PREVENTIVE MAINTENANCE

This system is normally used on critical equipment and includes in addition to the items listed in Minor Preventive Maintenance, periodic parts replacements and periodic overhauls.

PREVENTIVE MAINTENANCE DESIGNED FOR ZERO FAILURES

This system uses in addition to the items listed in major preventive maintenance, preventive maintenance instruction optimization, Analysis of Critical Problems (see chapter 11) and the Principal of Simultaneous Downtime (see chapter 12). Two examples of equipment falling under this category from the author's personal experience, are all facilities designated launch critical at Kennedy Space Center and two 36′ diameter × 300′ long rotary kilns in a cement plant at a time when the plant was operating 24 hours per day, 7 days per week and when all of the plant's output could be sold. (Downtime would therefore, result in an irretrievable loss in sales dollars).

CASE STUDY 1G: The Launch Critical Diesel Generator Set

At one plant, we had to retrofit 6 diesel – generator sets to provide power with tight specifications on voltage and frequency fluctuations and an allowable outage time of less than one second. The utility company power could not meet these specifications. The power was for critical computers and deviation from the specifications would cause a long list of catastrophic occurrences. As was the case with all critical equipment, we designed the preventive maintenance and operating systems for zero failure. Management, as usual, was willing to forgive one failure, but the second one was not permitted with the threat of being fired if the second one occurred. On one particular night, the operator had just shut down one set because it was making threatening noises; started up a spare; and was going through the sychronizing procedure. When he closed the circuit breaker, it blew up in his face and kicked all generators off the line. The previously mentioned catastrophies occurred and we worked furiously to get the set back on the line to minimize the delay. The problem was that we were forced to start up the generators without knowing what caused the explosion in the first place. After frantically investigating the mishap for one week, we were no closer to the solution and were afraid that the same thing would happen again. Although the operator had sworn that he had properly followed the paralleling procedures, it was concluded that the only possible answer left was operator error. When the operator was taken aside privately and offered immunity from disciplinary action, he confessed and the problem was solved. This incident taught two valuable lessons. These are:

- When a serious mishap occurs, there is no excuse for allowing a recurrence. An investigation should be started immediately with the goal of finding the cause and preventing a recurrence. (See chapter 11, Analysis of Critical Problems).
- A problem knows no organizational lines. The problem could be in any department, not only maintenance, and may not be exclusively a technical problem. It could, for example, be administrative, communication, personnel or a morale problem. In this case, the problem was with the operating department and was a personnel problem.

Part 2

THE DOWNTIME IMPROVEMENT CYCLE

8

General

Several things must be understood about the relationship between downtime and production lost costs, i.e.:

- Downtime does not always cause a production loss. (See Chapter 5, Criticality Determination for reasons why this is so.)
- Downtime on production equipment may be caused by items other than equipment breakdown i.e.:
    - Lack of tooling
    - Lack of operators
    - Lack of raw materials
    - Operator idle time
    - Lack of orders
    - Etc.
- Downtime on equipment chargeable to maintenance may be due to several causes (see Figure 2-1).
- Downtime should be recorded and charged to the department at fault, i.e: Maintenance, Production, Tooling, Material, etc.
- Even if downtime does not result in a production loss cost, it usually causes a loss in pro-

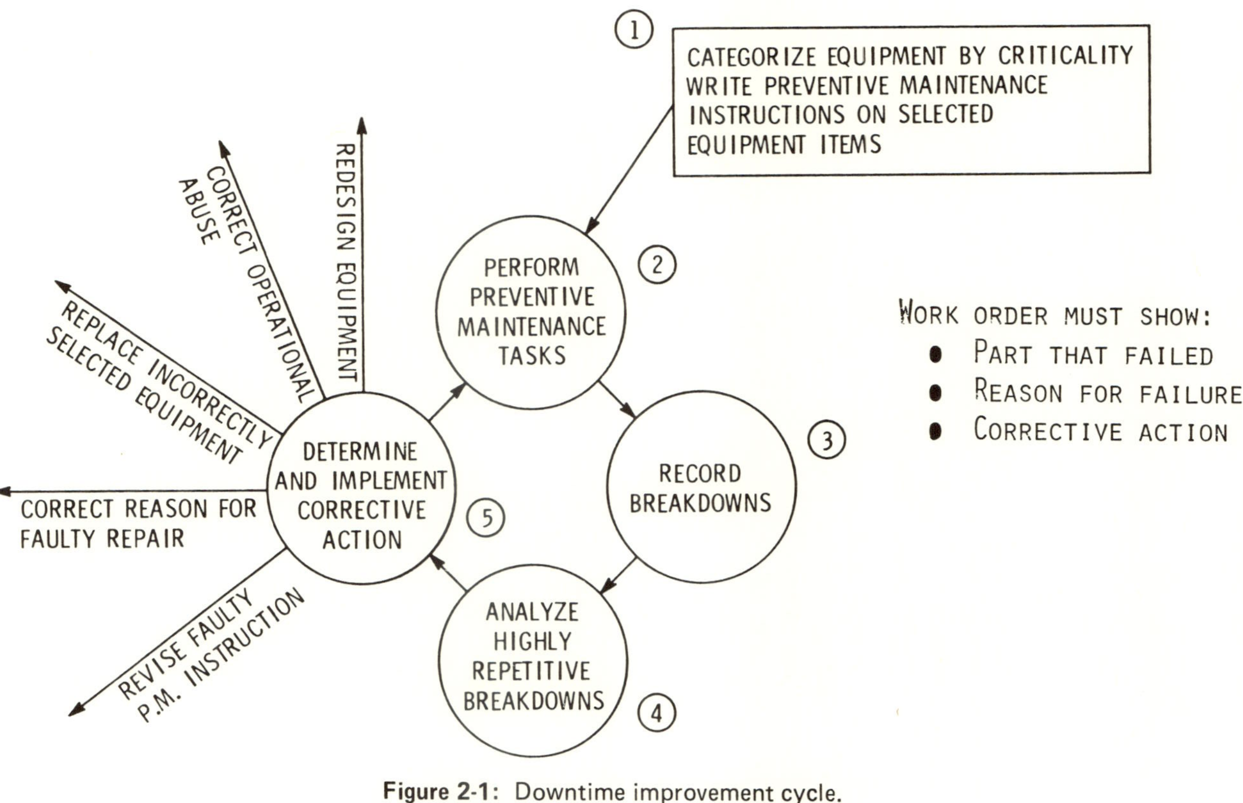

Figure 2-1: Downtime improvement cycle.

ductivity because people may be idled by the breakdown.

Downtime control can be achieved through a repeating Downtime Improvement Cycle (DIC), an acronym created by the author. This cycle is shown graphically in Figure 2-1. Downtime is reduced each time the cycle is repeated. This cycle consists of the following steps:

- Categorizing equipment by criticality and writing preventive maintenance instructions appropriate to the criticality.
- Performing preventive maintenance.
- Recording equipment breakdowns. Breakdowns should be sorted by equipment part that failed, reason for failure and corrective action taken to reveal highly repetitive and significant failures.
- Analyzing highly repetitive failures.
- Determining and taking corrective action. Such action usually falls in one or more of the following categories: redesign equipment; correct operational abuse; replace incorrectly selected equipment; correct reason for previous faulty repair; and revise preventive maintenance instruction. The last item closes the loop in the cycle.

The pages that follow will discuss the various steps in the downtime improvement cycle, starting with the recording of breakdowns. The reason for starting at this step is that this is the first system one would design and implement in an existing plant having no part of the downtime improvement cycle implemented.

9

Repair Records

GENERAL

The work order form is *the* document used to record equipment maintenance tasks and provides *the* most important source of information for maintenance analysis and the writing of PM instructions. In order to show how the work order is used as a maintenance record, a specific work order form that was designed is described, fully recognizing that each company will have its own particular needs which may dictate a different type of form. Particular emphasis is given to the work performed box, which is where the repair record goes, and the failure code, which is the device for entering this information into the computer.

WORK ORDER TYPES AND USE

The basic purposes of the Work Order System are as follows:

1. To provide a means for screening and authorizing work.
2. To provide cost data segregated in a logical manner.
3. To provide feedback information on repetitive failures for analysis purposes.
4. To provide a tool to facilitate planning and scheduling of maintenance work.
5. To facilitate control of productivity.

A typical work order form is shown in Figure 2-2. This form is used in two ways, as a *Blanket Work Order* and as a *Special Work Order*. The two ways that the work order is used are described below.

Blanket Work Orders: The concept of blanket work orders is used for two purposes.

1. On routine repetitive small jobs, when the cost of processing an individual work order may exceed the cost of the job. For example, a trouble call calling for the replacement of a single fluorescent light tube. A blanket order to cover these tasks might read: "To cover the cost of replacing light bulbs in the administration building".

The blanket work orders are normally written only once per year. When a man does work under this order, he merely charges the work order for the time he spent at the end of the day, thereby saving the time and cost of filling out and processing a special work order.

2. When a job is a fixed routine, such as a janitor's job. In this case, the janitor does the same thing every day. His job is repetitive and pre-planned. You know what the janitor does beforehand and the reporting of what he does on individual work orders by task would result in useless and costly information.

Special Work Orders: A special work order is written for all other individual jobs. Implied in the writing of a special work order is the fact that the particular individual job is important enough to warrant separate individual approval and reporting of all pertinent facts about the job.

The following paragraphs contain a description of the information that is to be put in each box of the work order.

Work Requested

Insert a description of work requested. The description should be as explanatory as possible. For example, if a production fore-

WORK ORDER NO.

WORK REQUESTED

WORK PERFORMED

JOB TITLE

COMPLETION APPVD: DATE:

MACHINE NUMBER	
MACHINE DESCRIPTION	
CC REQ.	CC PERF.
REQUESTED BY	
REQUEST DATE	PRIORITY
WORK TYPE CODE	

EST VS. ACT.				
CREW SIZE			HOURS	
EST	ACT		EST	ACT
		R		
		W		
		E		
TOTAL HOURS EST			ACT	

FAILURE CODE

Figure 2-2: Work order form.

man noticed something wrong with a pump, this box might read "Please fix bad leak in pump shaft seal". If the job was a pre-planned job written by the Chief-Planner Scheduler, the box might read "Perform complete overhaul on pump. Renew seals, all gaskets, broken nuts and bolts, clean thoroughly inside and outside. Renew gear; if necessary, repaint".

Machine Number: Enter machine number.

Machine Description: Enter machine description, such as "Storage Hall Crane".

Cost Center: Enter appropriate cost center numbers of the department requesting and performing the work.

Requested By: Enter signature of person requesting work.

Request Date: Enter date of request.

Work Order Number: Enter work order number.

Priority Code

The priority number is of major importance. It indicates about how much time is left before a repair need becomes a serious problem. A repair need becomes a serious problem when it:

1. is likely to stop production;
2. is likely to injure someone; and
3. is likely to damage equipment.

If any of the three conditions are likely to occur momentarily, you have an *emergency* condition (Priority Number 1). If you must interrupt your weekly schedule, but not your daily schedule, to avoid the possibility of any of the three above conditions, you have an *urgent* condition (Priority Number 2). If you are able to schedule the work at least one week in advance, you have a *normal* condition (Priority Number 3). If you have one month's lead time, you have a *programmed* condition (Priority Number 4). *Fill-in* work has no current time requirement (Priority Number 5).

Priority Numbers

No.	Name	Criterion	Action
1	Emergency	Production will be stopped unless repaired immediately. An extremely hazardous condition exists. Equipment will be damaged unless repaired immediately.	Repair immediately. Paper work to follow.
2	Urgent	A serious safety hazard exists and must be repaired before the end of the week and on the next shift, if possible. Production will be stopped unless repaired before next week.	Interrupt weekly schedule and place on next available daily schedule.
3	Normal	A defective condition has been identified. This condition will most likely not stop production, cause damage, or injure someone if corrected during the next week to four weeks.	Interrupt monthly schedule and place on next available weekly schedule.
4	Programmed	Predetermined repetitive repairs, period inspections, major maintenance repairs and construction will normally have this priority.	Place on next available monthly schedule.
5	Fill-in	Work assigned this priority has little or no time requirement.	Complete as time permits.

The priority system plays a vital role in communicating urgency in establishing criterion for manpower allocations and workload balancing. The Chief-Planner Scheduler shall enter the appropriate priority code in the box provided on the work order.

Estimate

The Chief Planner Scheduler (CPS) shall enter a rough estimate of the labor required to do the job in the space provided.

It is not intended that an exhaustive and very accurate estimate be attempted, but that a reasonable estimate be made based upon sound judgement. This will naturally require that the Chief Planner Scheduler be a person of considerable experience in all types of work.

Great care shall be made in selecting the most efficient crew size for the job because of the great effect this has on labor efficiency.

A detailed discussion on estimating methods and the reasons for and use of estimates is made in Part 3, "Productivity Control".

As can be seen from the estimating box on the work order, two quantities are required for each trade used, crew size and hours required per crew, i.e:

2 R 4

This means 2 repairmen are needed for 4 hours.

Job Title

The Chief Planner Scheduler shall enter a short accurate description of the job for input into the computer after the job is complete.

The reason for this step is to provide a meaningful job title in the computer reports.

For example, it is possible that the work requested box might read "Please fix slurry pump, it doesn't work". This is not a meaningful title. A meaningful title in this case might be "Replace motor overload elements", since it describes the work actually performed.

Work Performed

The foreman, upon completion of the job shall enter the actual work performed, including:

- Part that failed
- What corrective steps were taken?
- What caused the failure?
- What preventive steps should be or were taken?

Because the failure of one part frequently causes a chain reaction of failures, the part failing first must be identified. The information put in this box will vary depending on the circumstances, i.e.

Case 1 - Preplanned Work: When a job is preplanned in advance, for example, on a construction or modification job or on a planned overhaul; the work is completely described in the "Work Requested" box. The "Work Performed" box shall be left blank if the job is performed according to plan

Case 2 - Preplanned Work - Changed: If in the course of performing a preplanned job the work is modified in process, then a description of the change shall be entered in the "Work Performed" box. For example:

Work Requested

Install a 50′ – 0′ saddle in #7 Belt.

Work Performed

Job cancelled due to production problems.

Repaired damaged splice No. XYZ.

Case 3 - Breakdown Job or Trouble Call: In the case of a breakdown job or a trouble call, the work requested is frequently, of necessity, vague, i.e:

Work Requested

Fan #XYZ is vibrating badly, please repair.

After the fan is inspected and the fault isolated and repaired, the foreman knows the detailed cause and what was done.

Since this represents valuable feedback information for use in maintenance analysis, it must be recorded. it should be recorded in the work performed box as follows:

Work Performed

South Main bearing on Fan XYZ
dry of oil and burned up.

Replaced bearing

The recording of accurate feedback information on failures is one of the *most important* functions of maintenance foremen. You can depend upon the fact that good feedback will *normally not* be put on the work order without a strong campaign to achieve this by the maintenance manager, including lots of instruction and follow-up. The following is a typical letter that I wrote for the maintenance manager's signature to all of his foremen to emphasize the importance of adequate feedback:

TO: ALL MAINTENANCE FOREMEN

FROM: John Doe, Maintenance Manager

SUBJECT: INADEQUATE FEEDBACK DATA ON WORK ORDERS

An audit of completed work orders has revealed that the quality of information contained on the work order specifying the cause of failure is inadequate.

THE OBTAINING OF ACCURATE FEEDBACK INFORMATION ON THE CAUSE OF MAINTENANCE FAILURES IS ONE OF THE MOST IMPORTANT STEPS IN IMPROVING MAINTENANCE PERFORMANCE.

The information that we need and insist on getting on all work orders written for repair is:

(1) A description of the problem.
(2) A description of the part that failed.
(3) The foreman's judgement as to what caused the failure.

Item (1) is to be written on the work order under the heading "work requested." Items (2) and (3) are to be written on the work order under the "work performed" box.

An example will illustrate what constitutes adequate feedback data.

Trouble: Pump XYZ won't start.

Cause: Fuse blown. Fuse size was 10 A, should be 30 A. Replaced with 30 A fuse.

The above tells what the problem was; what part failed, and why.

The following are examples of unsatisfactory feedback data:

No Feedback on Work Order

Pump XYZ won't start.
(no cause recorded)

No Work Order Written

Pump failed, work charged to blanket work order, no written feedback.

Inadequate Feedback

Trouble: Pump XYZ won't start.

Cause: Replaced motor. (This does not state what the problem was.)

Trouble: Forklift inoperative.

Work Done: Replaced brake linings, replaced tires, ignition coil, spark plugs, changed oil, put in new solenoid valve and wiring. (This work does not specify the cause of failure.)

In the future all maintenance foremen shall record proper feedback on work orders in accordance with the principles outlined in this letter.

If a maintenance task is performed which constitutes valuable feedback information, the work shall not be charged to a blanket work order, since this would result in a loss of information. In this case, a separate work order should be written.

Concerning the cause of failure, it is recognized that certain failures will require an exhaustive and lengthy analysis. It is not expected that foremen will perform such an analysis, but should state, "It is suggested that Engineering investigate this problem." On the majority of jobs, however, it is anticipated that foremen will not only know the cause, but how to prevent a recurrence. In this regard, it should be noted that failures can be caused by:

(1) Lack of preventive maintenance
(2) Operator abuse
(3) Incorrectly selected equipment
(4) Faulty part design
(5) Poorly performed repair work, etc.

If the foreman isn't sure of the cause, he may be able to give his best guess of the cause. This should also be put in writing on the work order.

Completion Approved

When the job is completed, the foreman shall sign his name in the completion approved box. The signature will verify that he has inspected the completed work and that the work has been done in accordance with the plan (or modified plan) and that it is of proper quality.

Completion Date

After signing completion, the foreman shall enter the date completed.

Failure Code

A four digit failure code shall be entered after the job is complete and after *all* other entries are made. It is important that a person knowledgeable about the entire maintenance program make this entry since the information will be the main tool for maintenance failure analysis and a valuable tool for writing and updating PM instructions. A description of this coding system follows.

The system starts with a 4 digit failure code. The failure code is designed as follows:

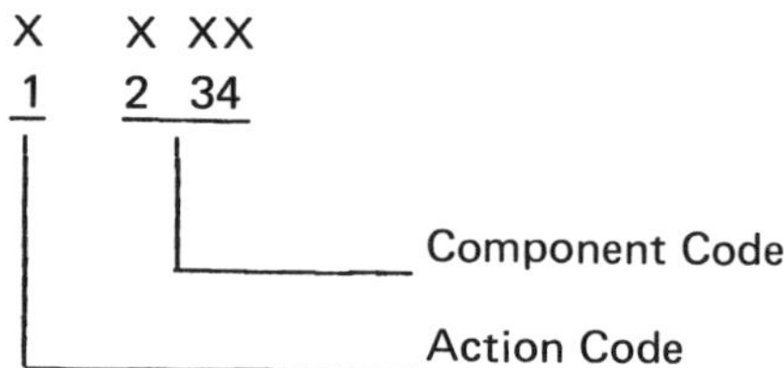

The definition of the Action Code is shown below. The definitions are self explanatory.

Code	Description
R	Replace
M	Repair
A	Adjust
C	Modification

These codes are a suggestion. Each maintenance organization may have need for more or different codes.

The last three digits are used to describe the component that failed. The following is a description of how to go about developing and updating this code:

Step 1: The code developer should call a meeting of all maintenance foremen and ask them to make a list of common component names used on equipment in their respective areas. For example, on a pump, the list might be:

Shaft
Impeller
Nut
Bolt
Packing
Impeller
Housing
Gaskets

This list alone includes components found frequently on many other equipment items.

Step 2: He then consolidates the list into one list eliminating duplications. This should include giving only one code to several component names meaning approximately the same thing (with suitable cross references). The more code numbers there are, the more difficult it will be to look one up and the more complex the failure report will be.

Step 3: Alphabetize the list.

Step 4: Number each component in sequence starting with 000 and going up to 999. This list will then be used by the work order coder for determining the last 3 digits of the failure code.

There will come the time when the person filling in the failure code will note that the component mentioned on the work order is not on the component list. He should then assign the next sequence number on the list to this new component.

This addition will result in the list being not 100% in alphabetical order and will therefore, make it increasingly difficult to find the item on the code sheet. This can be resolved by periodic submittal of the list to the computer department for alphabetical sequencing. This will result in there being two lists.

List 1: Sequential numbers, plus random component descriptions. This list will be used only for assigning new numbers.

List 2: Alphabetical component description list, plus random code numbers. This list will be used by the failure coder for finding the code number once the component name is known.

The purpose of putting the failure code on work orders will be described in the section on surfacing highly repetitive, critical failures in this chapter.

Work Type Code

A one digit code shall be put in this box to indicate the maintenance work type, i.e.

Code No.	Maintenance Work Type
	Maintenance
1	Major
2	Preventive
3	Routine
	Breakdown
4	Emergency
5	Planned
6	Trouble call
	Changes to facilities
7	Construction
8	Modification
9	Rearrangement
	Services
A	Janitorial
B	Waste handling
C	Landscaping
D	Other
	Maintenance Engineering
E	Planning and estimating
F	PM instruction writing
G	Maintenance analysis
H	Scheduling

It is very important to know at all times the percentage labor and material expenditures in each of the above categories, since there is a reasonable percentage for each company. For example, the ratio of breakdown to planned maintenance, is a measure of the success or failure of the preventive maintenance program.

CASE STUDY 2A: Percentage Construction Work?

A request was made to do a study of the maintenance department in a copper mine and plant which seemed to be suffering from a steady decline in performance over the past 3 years. Since there was no recording of maintenance work type on work orders, the coding system described above was immediately implemented. A check of the percentages based upon the first months trial run, revealed that 75% of the workers time was being spent on new construction.

A check of the company's history revealed that three years previous to the author's assignment, the company had become disenchanted with the performance of the firm that designed and constructed the plant. For two years they had been pushing the construction firm to clear up a long list of uncompleted an improperly designed facilities. In disgust, they finally threw the construction firm off the site and decided to complete the work on their own with the maintenance crew, which they considered more competent.

They had never dreamed that this task would tie up 75% of their maintenance crew.

If they had coded the work orders as described above, this fact would have surfaced immediately in the monthly computer runs, and they would have realized that they had "bitten off their nose to spite their face", and had made a disastrous error that had left inadequate numbers of maintenance workers to keep the plant running.

SURFACING HIGHLY REPETITIVE, CRITICAL FAILURES

Now that we have all of the information we need to analyze failures on the work order, all we have to do is extract the information from the work order in such a way as to pin-point the most critical and most highly repetitive failures. A computer program has been designed to do this very simply. The program makes use of the following information that is on the work order:

- Failure code
- Equipment number and name
- Criticality code
- Work order cost
- Date

With this information, the computer then prints the report shown in Figure 2-3. The report has the following features.

- Report can be programmed to print out only those equipment items that are the most critical (or any other degree of criticality)
- Equipment printouts are made in descending order according to the equipment item having the highest number of failures
- Failures are printed out in descending order of number of occurrences
- Number of occurrences this month and year to date total are shown
- Year to date cost is shown
- The failure record shows the action code, the component, the transaction date and cost, i.e:

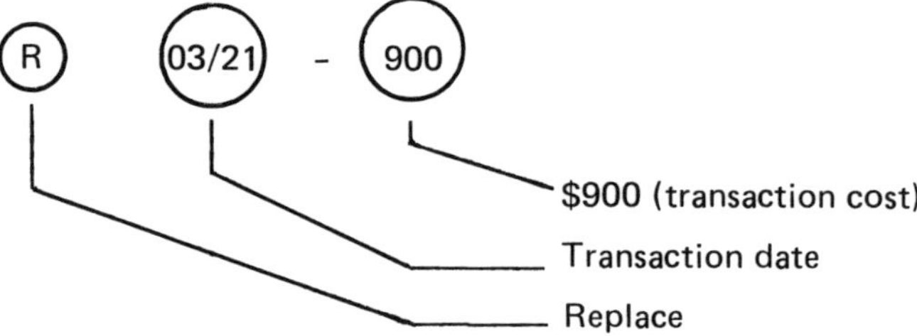

Thus, one can deduce from Figure 2-3, which is the first item on the printout that:

- Equipment has highest number of failures in plant.

December 1975	Historical Breakdown Record	580101 S. H. Crane		
Component	Failure Record (Type of Maintenance, Date, Cost)	Number of Occurrences		Year-to-Date Cost, Dollars
		This Month	Year-to-Date	
013 Holding Cables	M01/01-200, M01/03-210, M01/19-199, R02/03-1000, R03/21-900, R04/01-1100, R05/06-1000, R06/16-1090, R07/12-1500, R07/12-1500, M08/19-80, M08/21-99, M09/23-102, R09/16-990, R10/18-1000, R10/11-800, R11/03-900, R11/04-600, R11/15-500, M11/21-60, M11/26-94, M11/28-92, M12/01-50, M12/15-100	2	23	12,666
069 Brakes	R01/16-400, R10/11-350, M11/15-40	0	3	790
169 Bucket	M03/17-50, M10/21-75	0	2	125

Figure 2-3: Failure analysis report.

- Holding cable maintenance and replacement is the biggest problem with the S.H. Crane, by virtue of the high number of failures and cost to date.
- There were thirteen replacements of the holding cables in one year; an astonishingly high and unusual failure rate.

This high degree of visibility saves the maintenance analyst an enormous amount of work. Most of the time, the mere surfacing of a highly repetitive failure leads immediately to its remedy.

As stated previously, it is a good idea to simultaneously look at the tab run on parts usage listed in descending order of annual cost, as a cross check.

Only the December breakdown record reports should be filed (this is the annual report in effect). They should be filed in numerical sequence by equipment numbers.

Each month, maintenance will receive a monthly, year-to-date report consisting of historical records arranged in numerical sequence by equipment number. At this time, the previous months report can be thrown away.

The record system will therefore, consist of a monthly tab run discarded each month and a permanent file of annual reports for each equipment item.

Some plants contain many thousands of equipment items. This, coupled with hundreds of parts per equipment item, produces millions of components and a complexity factor capable of taxing the most competent plant engineer. He needs a tool to sort out the critical occurrence from trivia. The report shown in Figure 2-3 is such a device.

Assuming that a criticality code has been assigned to each equipment item, as described in chapter 5, the plant engineer has the option to order the computer to print out criticality code A (the highest criticality) equipment only. This could mean that only 100 out of 20,000 equipment items are reported on.

The report itself identifies which failure is significant by virtue of numbers of occurrences, total dollars, interval between failures and type of failure.

The maintenance analyst, after determining which failure is significant, refers to the specific work orders to obtain more details on the failure in an effort to eliminate or reduce the numbers of failures. Progress in correcting failures can easily be determined by examining the trends on the December (annual) reports.

When the maintenance analyst feels that he has solved all the problems in criticality code A equipment, he can program the computer to print out "B" equipment and "C" equipment thereafter depending upon the size and capacity of the maintenance engineering staff assigned to maintenance analysis.

PRODUCTION LOSS AND DOWNTIME COSTS

In chapter 6, the percentage that maintenance is of the total operating cost is discussed. If this percentage is high, the maintenance manager would be well advised to keep track of both downtime and production loss costs. As previously discussed, downtime can cost money by idling production workers, over and above the maintenance cost. Production loss costs may be suffered if a deadline machine causes a loss of sales.

Equipment, such as rotary kilns in a cement plant or a paper machine in a pulp and paper mill or a boiler in a utility plant, are so expensive to buy, and can cause such large losses when they break down, that a very complete record should be kept of all downtime occurrences, and the downtime or production loss costs should be charged to the department responsible for the downtime. Typical examples of the departments charged and causes are:

Tool department–Lack of tooling

Production department–Lack of operators

Material department–Lack of material

Maintenance department–Equipment failure

There are several ways to handle the accounting of these costs, i.e.

(a) Have the production operator keep the records right at the machine as a separate accounting system. The production department will then hold periodic meetings which show the percentage downtime by department and discuss ways to reduce the downtime.

(b) Have the department that suffers costs as a result of equipment downtime, charge the maintenance department for the cost. This can be done by them writing a maintenance work order and using codes 1 through 6 shown on page 46 for downtime costs or production costs. This system will isolate downtime and production loss costs and will put the pressure to reduce downtime where it belongs; on the maintenance manager.

If the machine is particularly critical, the maintenance manager may want to record downtime hours as well as costs. Foremen in this case would record the total amount of time that elapsed from the time a piece of equipment became inoperative because of an equipment failure to the time it was placed back into operation. This can be done after the job is complete in a special box labeled "downtime hours" on the work order.

This information will be used for two purposes.

(a) To plot curves of downtime trends for use by top management as a control tool.

(b) To provide additional information for foremen when analyzing repetitive failures.

When a foreman is analyzing a repetitive failure on an equipment item, he should be aware of the amount of time the equipment was deadlined. Obviously, it is a more serious matter if the equipment was out of service for four weeks as opposed to ten minutes. The equipment suffering four weeks downtime should generate in the foreman a greater sense of urgency in eliminating the repetitive failures. In the case of the four weeks downtime, not only should the foreman try to correct the repetitive problem, but he should look into the reason for the four weeks downtime. The reason may be material procurement time in which case he should work on shortening this time. Perhaps some of the repair parts were not stocked, or they may be stocked and the MIN-MAX is incorrect resulting in frequent stock outages. It is also possible that the repair time is excessive in which case he should consider repair methods improvement.

The following definition of terms will help in understanding the thorough analysis, which is sometimes necessary to make on critical equipment:

Total Downtime (TD)–Is the total time the equipment is out of service because it has broken down and is inoperable.

Production Lost Time (PLT)–Is that portion of the TD that production was lost.

Availability Time (AT)–Is the normal number of hours per week that the equipment is planned to be available for use.

Deadlined (DL)–Deadlined means that the equipment is broken down and unavailable for use.

Utilization Percentage (UP)–Equals AT − PLT ÷ AT.

The goal naturally, is to make *UP* as high as possible. TD is of value in surfacing the fact that the machine may be idled during a time when it is not needed for production, thereby, adding additional hours for repair.

The recording of production loss and downtime costs is necessary to optimize these costs with maintenance cost as described in chapter 4. In addition, it will be shown in chapter 13 that breakdown and preventive maintenance costs should be optimized regardless of whether or not there are downtime or production loss costs.

USE OF SUMMARY REPORTS

Summary reports can be of great use to the maintenance manager. The information for these reports is obtained from the work order system and is printed in useful form by the computer. Examples of useful summary reports are:

- Maintenance cost–$/Ton
- Maintenance cost–$/Unit produced
- Report shown in Figure 1-4.
- Janitorial costs in cents per square foot.
- Costs by work type code, including percentage each type is of total
- Parts use report arranged as in Figure 1-4

The reports should be subdivided by maintenance organization units. Standards and improvement goals should be set on the above indicators. Figures outside of standard should be further investigated by referring to the individual work order and in the case of failures, by referring to Figure 2-3 and then to the work order. Other examples of such reports follow.

(a) **Annual Equipment Report:** This report summarizes work order costs for the year by equipment number and is given in the following format:

Equipment No. xxxxxx

Work Order No.	Description	. . Labor . . (hr)	($)	Parts ($)	Outside Service ($)	Total ($)
x-xxxx	Replace drive motor	xxxx	xxxx	xxxx	xxxx	xxxx
	Totals	xxxx	xxxx	xxxx	xxxx	xxxx

This report should be printed in descending order of the total dollars column.

(b) Monthly Year-To-Date Major Problem Reports: The following two reports are recommended as a quick and easy means of surfacing principle maintenance problem areas:

Report #1

1 Equip. No.	2 Description	3 YTD Downtime Occurrences	4 YTD Maintenance Costs	5 Percent Column 4 to Total Column 4
xxxxxx	- - - -	xxxx	xxxx	xxxx

Report #2

Equip. No.	Description	YTD Maintenance Costs	YTD Downtime Occurrences	Percent Column 4 to Total Column 4
xxxxxx	- - - -	xxxx	xxxx	xxxx

All line items should be listed in decreasing magnitude of column #4.

The Preventive maintenance supervisor should be *particularly* on the lookout for equipment items that account for a *major* percentage of costs or downtime occurrences. For example, if the two reports showed that only ten equipment numbers accounted for 90% of the maintenance costs and downtime occurrences, then quite obviously these ten items should receive close to 90% of the maintenance analysis time. The intent is to prevent squandering of valuable analysis effort on an across-the-board effort with proper relationship between effort and return on effort.

It should also be emphasized that maintenance cost and downtime are *not* the only barometers of an effective maintenance program. There are many other factors that must be considered; for example, production loss and safety, to name a few.

(c) **Monthly Parts Usage Report by Equipment:** Another valuable tool for surfacing maintenance problems can be obtained by arranging the annual issues per stock item in descending order as follows:

			. . . Equipment Number:	XXXXXX . . .		
ACC #	Description	No. of Work Orders	Total Quantity Issued	Average Unit Price	Total ($)	Percentage
XXX1	Gears	X	X	X	3,000	67
XXX2	Rollers	X	X	X	1,000	22
XXX3	Bearings	X	X	X	460	10
XXX4	Rope	X	X	X	20	0.6
XXX5	Nuts	X	X	X	10	0.2
XXX6	Bolts	X	X	X	10	0.2
	Total				4,500	100

This report clearly shows that the major parts problem with this machine is *gears.*

10

Failure Analysis

GENERAL

The first step in failure analysis is to gather all of the information available that is written, and later on to supplement this with personal interviews with persons having knowledge of the failure. Written information that can be used is:

- Summary report
- Repair records (See chapter 9)
- Work orders
- Preventive maintenance instructions
- Correspondence files
- Purchase orders
- Stock records
- Analysis of critical problems
- Principal of simultaneous downtime studies
- Manufacturers data
    - Operation instructions
    - PM instructions
    - Trouble analyses charts
    - Spare parts lists

If an analysis of the above written data fails to solve the problem, then an interview should be conducted with one or all of the following persons who may have knowledge of the failure:

- Maintenance mechanics
- Machine operators
- Maintenance and production foreman
- Planner/schedulers
- Engineers.

USE OF THE FAILURE ANALYSIS CHART

The Failure Analysis Chart (Figure 2-3) is one device for isolation of equipment problems. Several observations are made relative to the use of this chart.

1. The depth of the column for each component failure gives an immediate visual indication of which failure occurs more frequently.
2. The frequency of occurrence is summarized in the table by month and year to date.
3. Another item which shows up is the time between occurrences. This must be weighed with the other facts. Something occurring ten times per month is obviously more serious than ten times per year.
4. A further indication of the magnitude of the problem is given in the year to date cost column.

After it has been determined that a problem is worthy of analysis by reason of the equipment criticality, the high repetitiveness of the failure or the high cost, a more detailed study must be made.

The first step is to obtain from the file all of the work orders listed on the Failure Analysis Chart for the particular failure being studied.

If the feedback on the work order is adequate, the detailed explanations on the work order should further isolate the cause of the failure.

Furthermore, the foreman's opinion of what the cause was will be recorded. If the feedback is inadequate, the analyzer will want to talk to the foreman and any other person knowledgeable of the problem and he will want to examine files containing drawings, correspondence, purchase orders, modifications, etc., on the equipment.

In the case of multiple failures, the analyzer must isolate the original part that set up the chain reaction (pyramiding failures).

DETERMINING THE REASON FOR FAILURE

Once a determination has been made as to which detailed part failed, the next question is why did it fail? The answer will be in one or more of the following general classifications (see Figure 2-4):

- Lack of preventive maintenance
- Improper frequency of preventive maintenance
- Faculty previous repair
- Incorrect equipment selection
- Faulty design
- Faulty manufacturer
- Operator abuse or error.

The biggest error that can be made at this stage is to say, "Oh well! It's just one of those things, parts will fail!" Or some similar statement indicating a lack of interest in the reason for failure.

Learning the *reason for failure* in detail is the "name of the game" in maintenance improvement. Foremen and the maintenance analyst must be indoctrinated in being inquisitive about the reason for failure on all breakdowns they are involved in.

Maintenance Analysis and Optimization functions should be directed towards achieving one of the following levels of maintenance effort:

ITEM	CAUSE OF FAILURE	CORRECTIVE ACTION
1	Faulty Design	Redesign
2	Faulty Manufacture	Warranty Action
3	Incorrect Equipment Selection	Modify Equipment Replace Equipment
4	Incorrect Operating Practices	Correct Operating Errors
5	No PM	Add PM
6	Incorrect PM	Correct PM
7	Faulty Breakdown Repair Work	Train Maintenance Craftsman

Figure 2-4: Cause of failures.

Highest level–Zero downtime

High level–Least possible downtime at the least cost

Normal level–Most economical maintenance point

Zero level–100% breakdown maintenance

The determination of which level to select is dependent upon factors discussed in Chapters 7 and 13.

It is a fact that the solution to the *majority* of repetitive failures is automatically found when the reason for failure is discovered.

For example, if a hydraulic or pneumatic hose system is failing much too frequently, the following questions might be asked.

1. What was the design pressure?
2. What was original hose and fitting specs?
3. What was actual pressure?
4. What was actual hose and fitting specs?
5. Were incorrect hose and fittings used?
6. Were they installed correctly?
7. Did the hose fail from overpressure, old age, erosion or corrosion from inside or outside?
8. Is the hose subjected to excessive vibration?
9. Is the hose subjected to continuous movement?

These are some of the questions that might be asked. The answer to these questions should lead to the reason for failure.

DETERMINING CORRECTION ACTION

As was stated previously, the majority of solutions to highly repetitive problems are found automatically the same time the reason for failure is determined. To demonstrate this, a few actual examples are given.

PROBLEM 1.

Frequent hydraulic hose failure

Cause:

Hose subjected to overpressure due to faulty operation of pressure reducing valve.

Solution:

Repair reducing valve. Install relief valve on downstream side set to protect hose against overpressure.

PROBLEM 2.

Frequent pump outages.

Cause:

Fuses blowing in control circuit. Fuse was incorrect size and type.

Solution:

Installed correct size and type of fuse.

PROBLEM 3.

Frequent hydraulic hose failure on *all* vehicular heavy equipment.

Cause:

Foreman has substituted 2-ply plain hose with manually installed fittings instead of original design which was 5-ply armoured hose with machine swedged fittings because new hose was easier and simpler to install.

Solution:

Purchase swedging machine. Use specified hose and fittings.

The solution to some problems may be above the skill or responsibility level of maintenance foreman either because of technical complexity, large dollar amounts required for the fix or because of the need for a process modification. These problems should be handled by the maintenance or engineering staff as directed by the maintenance manager. In some cases, the problem may not be solvable by the company's staff. If the consequences of the repetitive failure are serious enough by reason of cost and lost production, consideration may be given to the use of outside consultants.

There are several pitfalls that the analyst must guard against in his problem solving function.

1. The tendency of many analysts to "drop the ball" if *their* solution is turned down. When an idea is turned down, the following questions must be asked:
    - Was it turned down for cost or lack of funds?
    - Was it turned down because the approver didn't think it would work?
    - Can it be done cheaper?

- Can it be done differently?
- Can the scope of the solution be reduced to save cost?
- Was the savings justification adequate?

2. The tendency of some analysts to "drop the ball" if the fault lies outside of their department, i.e.: operational abuse, incorrect equipment, problem is electrical, or a training or personnel problem.

   Analysts must be prepared to cross organizational or functional lines when necessary to solve a problem.

EQUIPMENT OUTAGES AND RELATIONSHIP WITH PREVENTIVE MAINTENANCE

Remembering that our maintenance goal is to achieve the lowest combined cost of maintenance, plus production lost time, it is important for the maintenance analyst to carefully consider the question of equipment outages.

The manner in which equipment outages are handled is important for several reasons:

- It affects production loss cost
- It affects crew efficiency
- It affects the ratio of production loss to major overhaul cost.

The handling of equipment outages can be divided into two categories:

- Emergency Outages
- Preplanned Outages

Emergency Outages

The most important first step during an unanticipated breakdown is to arrive at a decision of what to do about the failure.

Run the equipment as is? Make a patch repair? Make a proper fix? Fix other things as long as the equipment is down?

The time required to make this decision causes production loss and a productivity loss for the maintenance crew. The decision of *how* to fix it also affects production loss. It is important that the decision be made *quickly* and *correctly.* this requires the collaboration of a knowledgeable person from maintenance and production. A perusal of hundreds of critical problem reports from other companies shows that errors in judgement concerning this subject are not uncommon.

The analyst should:

1. Discuss the subject with maintenance and production foremen periodically to sharpen their sensitivity to the problem.
2. Consider this subject in the writing of critical problem reports. Poor performance in this area surfaced by the critical problem report should be made the basis for corrective action including briefing of foremen on the problem and solution.

Preplanned Outages

Preplanned outages for maintenance on production equipment running intermittantly or with spares or in-process storage arrangements are no problem as far as production loss is concerned. It is the equipment that is running around the clock that is the problem.

There are several things which can and should be done to optimize the relationship between downtime for *preventive* maintenance purposes and production lost time cost. These are:

1. Identify equipment which may cause a production loss when inoperative.
2. Identify if equipment is producing parts into a stockpile and how long the stockpile will last. (This time may allow downtime without a production loss).

3. Record existing annual preventive maintenance program on the equipment and its statistical performance including:

    - Maintenance cost per ton or per $ of product output
    - Downtime hours per year
    - Downtime percentage

4. Determine if there is redundant equipment.

These four items represent what the present maintenance plan is and its performance. The next step is to outline several alternate plans and *estimate* their probable performance characteristics. Those plans which appear very good on paper may be worthy of an actual test. Alternative methods which may be considered are:

1. 100% Breakdown Maintenance

2. Light Preventive Maintenance Plan: This includes light preventive maintenance work on items that can be worked on while running or appurtenances that can be shutdown or that are provided with backup.

3. Heavy Preventive Maintenance Plan: This includes a major overhaul at some logical frequency which depends on the equipment; i.e., annual, semi-annual or bi-annual.

An example of this plan is the method of maintaining reciprocating aircraft engines. This plan includes certain preventive maintenance functions that must be performed after 100 hours, 500 hours and 3,000 hours of operation. The intention of this preventive maintenance program is to achieve zero downtime.

4. Operational Monitoring: The best example of this plan is the present method of maintaining jet engines on many of the top airlines. This plan includes the installing of instrumentation which measures the performance of the components under the plan. For example, the fuel pump might be instrumented to record fuel pressure, flow, bearing temperature and vibration. These measurements are continuously recorded and intermittantly transmitted to a central computer for analysis. If the operating characteristic exceeds the tolerances on the standard, an order is issued for a repair or replacement. In other words, no maintenance is ordered unless there is a deviation from operational standards and tolerances. It is reported that this plan has more than doubled the operating life of the engines and made a drastic reduction in maintenance cost.

5. Combination Plans: Finally, combination plans, using any or all of the above methods, may prove to be worthwhile.

Changes in the maintenance programs for very large, costly and critical equipment should be approached with caution because of the large costs involved. For this reason it is wise, if possible, to try and reason out the impact and results of the change on paper first. A well thought out "mock trial" may not be exactly similar to actual operation, but may be close enough to see if it is an obvious improvement or not.

6. An Example: An example of the above should help in illustrating the point:

Company

A mining firm.

Equipment

39–1000 HP synchronous motor drive ball mills.

Old Maintenance Plan

Variable frequency preventive maintenance

Existing Historical Records

Firm had very complete historical records on failures.

Existing Average Downtime

15%

This company in analyzing its records noted that there was a random failure pattern averaging three months. An analysis on paper showed a huge savings in downtime if the random failures could be consolidated to one downtime occurrence.

The "mock trial" calculations were made as follows:

Existing Operating Hours

39 mills × 8,736 hours per year = 340,704 hours per year

Proposed Downtime

Take 3 ball mills down for maintenance (plus appurtenances) for 8 hours once per week. Since there are 39 mills, it would take 39 ÷ 3 = 13 weeks to do this to all mills. Then the cycle would be repeated. This would result in each mill getting an 8 hour overhaul, 4 times per year.

During the mill outage all items would be open for inspection and faulty items fixed as far as is possible. If 8 hours was insufficient, the mills would be worked on around the clock until complete.

Downtime = 3 mills x 8 hr/day x 1 day/wk x 52 wk/yr =
1,248 mill hr/yr
Total mill operating hours =
39 mills x 24 hr/day x 7 days/wk x 52 wk/yr = 340,704
Downtime percentage = 1,248/340,704 x 100 = 0.4 of 1%

Results

The plan was put into effect and the downtime was reduced to 1% or only 0.6 of 1% over estimate. This was a reduction of 14% in downtime; a noteworthy improvement.

There was considerable resistance to this change since production didn't relish the idea of "unnecessary" planned downtime. Unnecessary being interpreted to mean "Why shut down if it is still running?"

As the years went by with this new system, they found that certain items would always need replacement or repair during the shutdown. They would be ready with parts needed for these items. They would also experiment with new designs or materials on highly repetitive failures on one mill. If the new design or material was an improvement, the improvement was put on all mills.

When there was an uncertainty about a certain part, the question was asked "Will it last another 3 months?" If the answer was yes, it was left alone and a note made to inspect it during the next overhaul. The result of this analysis was to reduce the planned downtime from 8 to 5 hours for each mill.

PRODUCT SPILLAGE

Product spillage, dust generation, litter and uncollected trash is another problem which may have unfavorable side effects not experienced in an ordinary equipment malfunction; i.e.:

Low Morale

Nobody likes to work in an unclean environment.

Production Loss

Product spillage may result in damage to equipment and eventually cause a shutdown.

Health Hazard

Uncollected fumes, dust from spillages, etc., may result in a hazard to health.

Safety Hazards

Product spillage may cause safety hazards.

Fire Hazards

Uncollected trash, papers, oily rags, or other combustibles may create a fire hazard.

Low Productivity

An unclean plant creates an atmosphere of disorderliness and confusion and may lead to low productivity.

Generates Further Uncleanliness

When a plant is not clean and neat, there is more of a tendency to make it worse. Nobody likes to throw a cigarette butt on a clean floor. However, if the floor is already a mess, nobody hesitates.

Product spillage and dust is just one of the manifestations of equipment malfunction and should, therefore, be handled in the same manner. This includes all phases of the downtime improvement cycle. Let us visualize a simple belt maintenance program for example:

Preventive Maintenance

None

Breakdown

Product spillage thru skirts

Analysis

Product is spilling thru worn skirts and building up in a pile which submerges the belt and rollers. Estimated monthly breakdown cost is $500 per month and 16 hours downtime.

Corrective Action

Write preventive maintenance instruction to change skirts and scrapers once per month. Estimated cost $150 per month.

In analyzing a spillage problem, consider the following:

1. What component failed and what can we do to prevent reoccurrence of the failure and its consequences? (Let us assume that the failure is the worn and torn skirt; the consequences are product spillage, damage to the belt and rollers and equipment outage).

2. Why didn't production personnel notice the spillage and at least push it out of the way far enough to prevent the equipment damage and downtime?
3. If the cause of the spillage is not going to be stopped immediately, what can be done to remove the spillage and prevent further equipment damage and downtime?

A spillage problem can be resolved either partially, or totally, by one or more of the above alternatives. The mistake that can be made in this respect is to try to solve the problem by means of one alternative and then to drop the entire investigation upon disapproval of this alternative without investigating the other alternatives.

IMPLEMENTING CORRECTIVE ACTION

It is anticipated that once the feedback systems are properly functioning together with the maintenance analysis function, a backlog of work will soon develop. The backlog of work should go through the following processing steps:

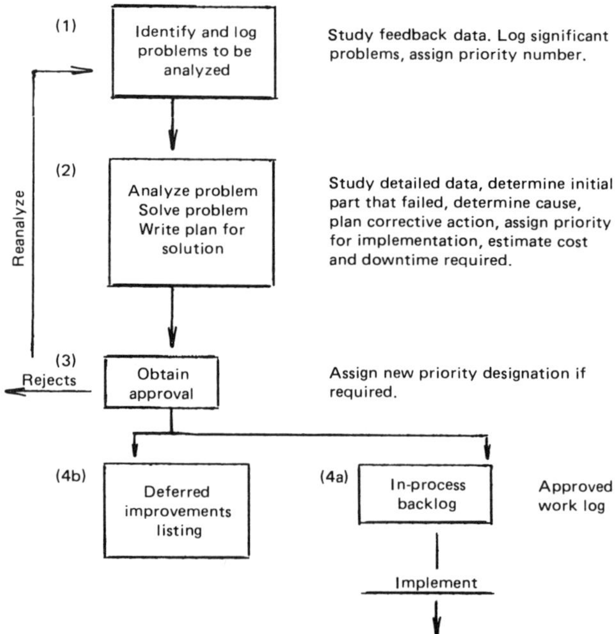

It is suggested that a log book be kept to record the magnitude of the analyzing workload and of the approved corrective plans. The desirability of the corrective effort can be prioritized by the assigning of a simple priority code as follows:

Code	Description
M	Solution is mandatory
HD	Solution is highly desirable
D	Solution is desirable

Solutions to problems which have a marginal ratio of benefits to cost can be either:

1. Rejected
2. Referred for further analysis
3. Logged in a deferred improvements log

An important part of implementation is the follow-up analysis to determine if anticipated benefits are achieved. Confirmation of benefits can be determined by watching the changes to the failure analysis chart each month. A complete elimination of the problem may be evidence of an "overkill". For example, a change from lubrication from once per year to once per month may eliminate a bearing problem, but may also cause unnecessary preventive maintenance costs. The correct period for any particular plant may have to be determined by trial and error.

MAINTENANCE ANALYSIS EXAMPLES

Examples of typical analyses are shown in Figures 2-5 to 2-9. The following comments are made relative to these examples:

(a) Note that production loss cost is included as part of the cost of the maintenance failure. This is a *very* significant point. The reason it is significant is because this cost can at times dwarf the pure maintenance cost. As a result it sometimes pays to spend *more* money on maintenance if it will reduce the production loss cost and, therefore, the total cost of maintenance.

Maintenance Analysis Form			
Historical Record: *Truck No. XYZ* *Clutch being replaced twice per year*			
Old Method: *Operator was riding the clutch.*	**New Method:** *Train operator in correct operating method.*		
Item Description	**Cost in Dollars**		
	Old	**New**	**Diff.**
Labor: *Replace clutch $100 x 2*	*200*	*0*	*200*
(Assume car will be traded at 60,000 miles and that no clutch replacement will be required.)			
Material:			
Clutch $50 x 2	*100*	*0*	*100*
Production Loss Cost: *None*			
Total:	*300*	*0*	*300*
Downtime in Hours	*32*	*0*	*32*

Figure 2-5: Maintenance analysis example 1—incorrect operating practice.

(b) When comparing an improved method with the old technique, *it is usually not necessary* to make the comparison in writing as shown in the examples. Also, it is not necessary to be extremely accurate with the estimate. "Guesstimates" will be sufficient. The purpose of the examples is to show the thinking process.

Maintenance Analysis Form

Historical Record: *Truck 2 PT*
Tires are replaced quarterly. Annual mileage is 24,000 miles.

Old Method:	**New Method:**
No Tire PM	*Worn tires are being caused by underinflation. Instructed operators to check tire pressures daily.*

Item Description	Cost in Dollars Old	New	Diff.
Labor: *Old*			
4 tires x 4 times/year = 16 tires x 10 min/tire			
= 160 min or 2.7 hrs			
2.7 hrs x $4/hr	*10.80*		
New			
$\frac{10.80}{4}$ *= 2.70/yr*		*2.70*	*8.10*
Material: *Old 16 tires @ $100.00 each*	*1,600*		
New 4 tires @ $100.00 each		*400*	*1,200*
Production Loss Cost: *None* -			
Total:	*1,610*	*403*	*1,202*
Downtime in Hours	*8*	*2*	*6*

Figure 2-6: Maintenance analysis example 2—no preventive maintenance.

(c) In cases where the old and new methods compare very closely, the comparison should be done in writing.

(d) All of the comparisons given in the examples are clear cut as far as whether the recommendation (New Method) is better. This is not always true, nor are the costs of a new method always known, nor are the consequences of a new method always predictable.

Maintenance Analysis Form

Historical Record: *Equipment No. XYZ*
Main fuses failing at rate of 10 per month.

Old Method:	**New Method:**
Use 10 amp fuses	*Investigation revealed design error. Fuses should be sized at 20 amps.*

Item Description	Cost in Dollars		
	Old	New	Diff.
Labor: *Travel time to and from 20 min.*			
Diagnose trouble 5 min.			
Replace fuse 5 min.			
Total 30 min.			
@ $4.05/hr = $2.00			
10 occurrences x $2	20	0	20
Material: *Fuse – 10 amp, 10 x 50¢ = $5*	5	0	5
Production Loss Cost: *None – In process storage and redundant units available*			
Total:	25	0	25
Downtime in Hours	10	0	10

Figure 2-7: Maintenance analysis example 3–faulty design.

INDETERMINANT MAINTENANCE

As mentioned in the last paragraph, it is not always possible to predict exactly what will happen when a change is made in a maintenance task. For example, what will happen if you change the greasing of a motor from once per three months to once per four months?

Maintenance Analysis Form			
Historical Record: *Floor Sweepers (1.0 of)* *Drive belts are failing on the average of once per 10 months*			
Old Method: *PM instructions on 10 floor sweepers calls for drive belt inspection once per month.*	**New Method:** *Belt inspection once per month uneconomical. Discontinue inspection and call for belt renewal semi-annually (since belts are now failing @ 10 months.)*		
	Cost in Dollars		
Item Description	Old	New	Diff.
Labor: *Old*			
Inspection: 10 units x ½ hr per x $4/hr x 12 times per year	*240*		
Belt Replacement: 10 units x $4/hr x ½ hr/unit x 12/10	*24*		
New			
Belt Replacement – 2/yr x 10 units x ½ hr/unit x $4/hr		*40*	*224*
Material: *Old*			
Belts – 10 units x $5/set x 12/10 yr	*60*		
New			
10 units x $5/set x 2/yr		*100*	*-40*
Production Loss Cost:			
None			
Total:	*324*	*140*	*184*
Downtime in Hours	*0*	*0*	*0*

Figure 2-8: Maintenance analysis example 4—incorrect PM.

You think nothing different will happen, but you're not sure. In other words, the consequence of the change is *indeterminant.*

This is frequently the case when you are contemplating a change in frequencies. The only thing that can be done under this circumstance is to guess, and to determine action by this "guesstimate".

Maintenance Analysis Form			
Historical Record: *Conveyor XYZ (Category A)* *Skirts and scrapers are being worn out and replaced 6 times per year.*			
Old Method: *Belt fasteners installed without countersinking of belt. The raised fastener is tearing skirts and scrapers.*	**New Method:** *Instruct craftsman to countersink all belt fasteners.*		
Item Description	**Cost in Dollars**		
	Old	**New**	**Diff.**
Labor: *Old*			
6 occurrences			
4 hours/occurrence			
$4/hour	*96*		
New			
1 occurrence per year		*16*	*·80*
Material: *Old 6 occurrences*			
$40 for skirts and scrapers			
$20 for fasteners	*360*		
New 1 occurrence per year		*60*	*300*
Production Loss Cost: *Prod. loss cost is $1,000/hr on critical belt.*	*6,000*	*1,000*	*5,000*
Total:	*6,456*	*1,076*	*5,380*
Downtime in Hours	*6*	*1*	*5*

Figure 2-9: Maintenance analysis example 5—faulty repair work.

The following example will help to illustrate the "guessing" procedure (See Figure 2-10):

> Let us assume that we are suspicious of excessive bearing failures on equipment XYZ, and, furthermore, we suspect that the oiling frequencies are incorrect. We proceed to "guess" at

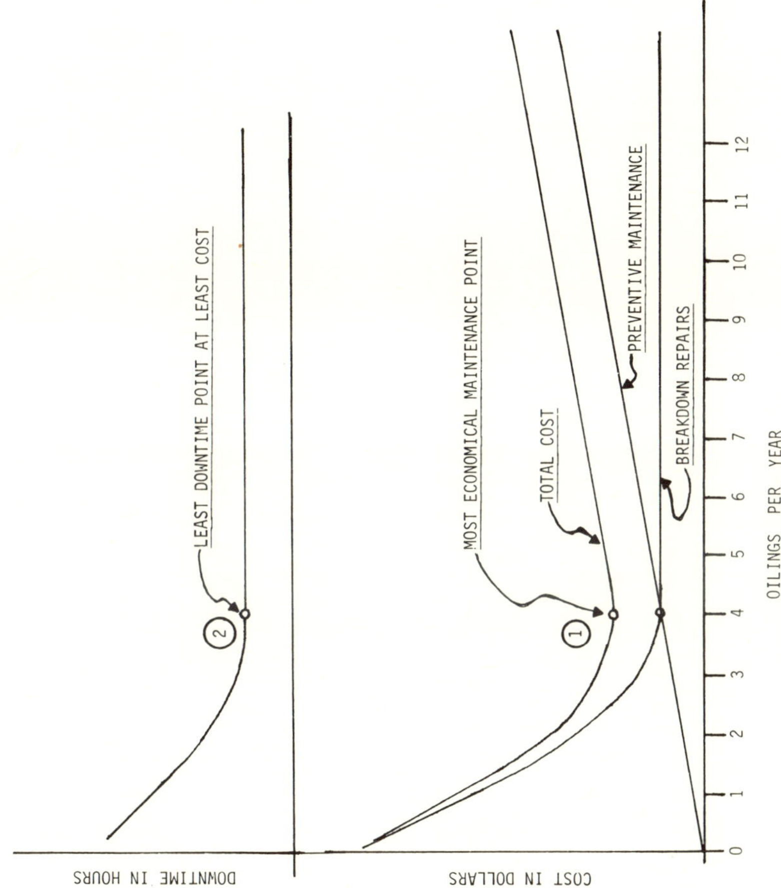

Figure 2-10: Maintenance analysis—guessing procedure.

> what would happen if we were to change these frequencies.
>
> Concerning preventive maintenance cost, we can assume that its cost will be a straight line function. In other words, every time we increase the frequency by one, the preventive maintenance cost will go up by one cost unit.
>
> Breakdown repairs are of course, astronomical with zero oilings. However, this cost drops sharply as oiling is started. We guess that bearing failures will be negligible when we reach a frequency of four times per year.
>
> When the Preventive maintenance Cost Curve and the Breakdown Cost Curve are added together, we get the most economical maintenance point (1).
>
> We also conclude that the downtime hours will stop diminishing when the oiling frequencies go beyond points (1) and (2).
>
> Corroboration of our "guesstimate" will take place by watching the equipment records. In this case, our "guesstimate" leads us to the conclusion that the bearings should be oiled four times per year.
>
> The understanding of *how* to arrive at points (1) and (2) is *very* important since these points represent the two desired goals of a preventive maintenance system. The points are arrived at by TRIAL AND ERROR! At times, there is no other way.

Once again, it is important to understand that it is not necessary nor desirable to fill out the forms or plot the graph used in this section to illustrate the thought process. But, it is necessary to think in the manner described.

11

Analysis of Critical Problems

GENERAL

From time to time, problems occur that are of such a serious nature that a formal and immediate investigation is justified. Furthermore, analysis and corrective action performed at top management level is considered mandatory. Critical problems may fall within the following areas:

(a) Maintenance failures causing serious production stoppage.

(b) Operational mishaps causing serious production stoppage.

(c) Maintenance failures or operational mishaps which *may* cause serious production stoppage.

(d) Maintenance failures or operational mishaps which caused an accident or potential serious safety hazard.

The purpose of formalizing the investigation is to insure that the problem receives a thorough and penetrating analysis designed to prevent problem reoccurrence. Nothing is more embarassing or inexcusable than to have repetition of a serious problem because the previous problem analysis was:

(a) Not made
(b) Sloppy
(c) Incomplete
(d) Offered no solution.

The author has been repeatedly shocked as a maintenance consultant to find, in many of the plants that he surveyed, such a large number of critical events that were allowed to pass by without an investigation. As a matter of fact, many of these serious events occurred continuously over a period of many years. Plant personnel considered the event "just one of those problems that you had to put up with" in running their type of plant. In this section a procedure to use in order to insure that critical problems are investigated is outlined, and then at the end, several examples are given out of the author's experience to illustrate the procedure. (The actual procedure may vary with each plant.)

REASONS FOR UNSATISFACTORY ANALYSIS OF CRITICAL PROBLEMS

Before a description of the required report is made, it may be helpful to describe some of the reasons why unsatisfactory analysis of problems is sometimes made. Practically all of these reasons can be traced to one person not assuming total responsibility for the entire analysis:

(a) Production foreman ignores maintenance problem because it is not his responsibility.
(b) Maintenance foreman ignores production problem because it is not his responsibility.
(c) Mechanical foreman ignores electrical problem because he knows nothing about electrical work.
(d) Person investigating problem takes someone else's word about some detail without making his own study because of lack of knowledge of subject matter or hesitance to cross organizational lines.
(e) "It's just one of those things that happen." "Why get upset about the inevitable?"

The point is that problems cross organization and technical expertise lines. It is up to the problem analyzer to cross these lines and make a thorough analysis regardless of his reporting position or technical skill.

First line foremen are charged with the responsibility of promptly notifying management when critical problems occur.

DESIGNATION OF CRITICAL PROBLEMS

After being notified that an incident has occurred, the vice president of operations, or equivalent position, should determine whether or not to classify the incident as a critical problem.

RESPONSIBILITY FOR ANALYSIS

The maintenance manager, plant manager, or vice president of operations should designate the *single* person responsible for making out the Critical Problem Report. All reports, however, should be signed by one of the above three persons, indicating top management interest in having the problem solved. The writer of the report should assume *full* responsibility for the worth of the analysis and should make a presentation of the report to top management.

REPORT DUE DATE

The report should be submitted in stages as follows:

(a) Verbal Report

Due within the hour to the maintenance manager. The manager should verbally inform the plant manager and vice president of operations immediately thereafter.

(b) Final Report

A final written report is due on the desk of the maintenance manager, the plant manager and vice president of operations within 24 hours of the incident.

REPORT FORMAT

The report shall employ the following format:

Critical Problem Report

No.________ Date________ By________

1. What happened?
2. What corrective action was taken?
3. What was the impact on production?
4. What was the cause?
5. What will be done to prevent recurrence?

The "number" listed above shall be an eight digit number determined as follows:

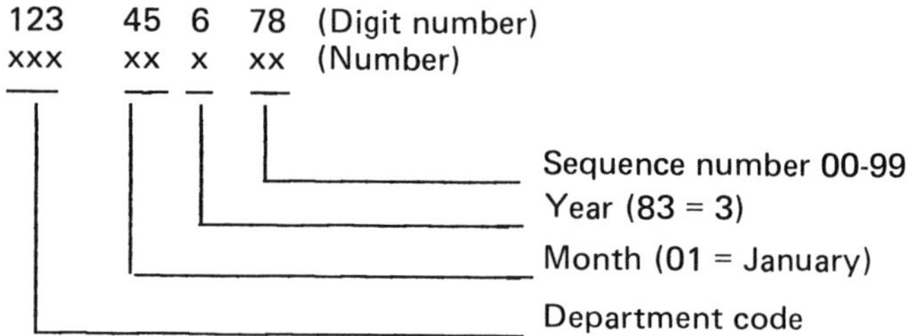

WHAT HAPPENED?

Explain what happened including what, when (date, hour) and how. Be complete, name who and what was involved.

WHAT WAS IMPACT ON PRODUCTION?

What was production downtime hours? Completely describe impact. What else was affected?

WHAT WAS CAUSE?

What was cause of problem?

— Lack of preventive maintenance
— Operator error
— Operator carelessness
— Faulty manufacture
— Faulty installation
— Act of God
— Design error
— Etc.

Go into sufficient detail to surface specific cause as opposed to general cause.

WHAT CORRECTIVE ACTION WAS TAKEN?

What was done to correct the problem? Was the fix temporary or permanent?

WHAT WILL BE DONE TO PREVENT RECURRENCE?

How will you guarantee that the event does not reoccur?

If you are not sure of how to prevent reoccurrence, state so and describe recommended investigation and research effort required to arrive at the final solution.

CRITICAL PROBLEM LOG

The secretary to the plant manager or vice president of operations should maintain status of all critical problems in a log book. Status shall be obtained weekly from the maintenance manager on unresolved problems.

A sample log book sheet is shown in Figure 2–11.

Critical Problem Log																						
Number	Date of Occurrence	Description	Status Code																			
			DEPT 1					DEPT 2					DEPT 3					DEPT 4				
			1	2	3	4		1	2	3	4		1	2	3	4		1	2	3	4	

Figure 2-11: Critical problem log.

The age of the problem will be shown in the log book. The secretary should list the problems in descending order of age. This will provide increasing pressure on the maintenance manager to resolve the problem as its age increases. As the problems are successfully resolved, they can be transferred to a "Resolved Critical Problem Log", which will be in a sense, a log of progress.

I suggest using only one of the following phrases in the status column:

Status	Code
Unresolved	1
Temporary fix made	2
Permanent fix design only complete	3
Permanent fix implemented	4

DISCUSSION MEETINGS

Meetings should be called by the plant manager or vice president of operations, as necessary, to update and satisfy him on the progress towards solutions to outstanding problems.

EXAMPLE

In order to illustrate the proper filling out of a critical problem report, the following example is offered:

CASE STUDY 2B: The Fuel Pump

Critical Problem Report

No. 511–113–04 Date 11/24/82 By John Jones

1. What Happened?

8 P.M. (Nov. 11)

Fractionating tower operator Smith reported over temperature condition on south main fuel pump bearing to control room.

9 P.M.

Operator Smith reports outage on main fuel pump.

9:10 P.M.

Auxiliary fuel pump started manually. Tower capacity reduced 30% to compensate for smaller size auxiliary fuel pump.

9:15 P.M.

Shift maintenance supervisor, Dave Doe, notified. Doe decides to have repair started on day shift due to existing heavy workload.

7 A.M. (Nov. 12)

Maintenance manager notified of incident at home by midnight foreman.

7:15 A.M.

Maintenance manager notified plant manager of incident.

8 A.M.

Dismantling of damaged fuel pump initiated.

10 A.M.

Dismantling completed. Assessment of damage as follows:

South bearing seized, melted, totally destroyed

North bearing damaged

Shaft broken

Flexible coupling damaged

Motor tripped on overload. Motor OK

Impeller badly damaged

(Pictures of the damaged pump are attached herewith)

2. What Corrective Action Was Taken?

11 A.M.

Machining of new shaft started. Shaft completed 3 P.M.

12 Noon

Bearings confirmed as stock item. Item out of stock. Flexible coupling found in stock. New impeller located in stock.

3 P.M.

Purchase order complete. Bearings ordered air freight by phone.

9 A.M. (Nov. 13)

Bearings arrived in plant in receiving. Maintenance supervisor notified of delivery 10 A.M.

12 Noon (Nov. 13)

Pump rebuilt, reinstalled and put into service.

3. What Was Impact on Production?

Outage time:	9 pm November 11 to 12 noon, November 13	39 hr
Production loss:	39 hr x $5,000/hr x 0.3 =	$58,000

4. What Was the Cause?

Bearing was a manually oiled type requiring daily oiling. Regular oiler off on vacation. Replacement oiler neglected to oil bearing. Bearing wasn't oiled for two weeks. Although there was a written oiling check sheet, foreman neglected to review sheet or give sheet to replacement oiler. Replacement oiler had filled in on this job several times previously and was thought to be familiar with assignment. Oiler interviewed and stated since all other similar pumps had greased bearings, he assumed this pump had greased bearings.

5. What Will Be Done to Prevent Reoccurrence?

This failure was unjustified. Furthermore, the length of the outage was found to be excessive as a result of a detailed analysis of the incident. The following is an analysis of the incident broken down by technical, preventive maintenance and communication aspects.

Technical Problem

The cost of this outage alone would justify a full sized standby fuel pump electrically interlocked to start up upon failure of the main pump pressure.

The original design of the fuel pump calls for greased ball bearings. Someone made a design modification on this pump altering the bearing type. No record could be found of this modification or of previous bearing failures.

Furthermore, no maintenace supervisor recalls the modification. This mod is in violation of established procedures calling for the chief engineer to sign off on equipment mods.

Preventive Maintenance Problem

The oilers are in possession of daily check off sheets for this oiling route. Investigation of procedures reveals that the sheets haven't been used for at least two years. The present oiler has been on this assignment for one year. The maintenance foreman in charge of the oilers doesn't recall when the practice of using the sheets was discontinued. The Maintenance superintendent says that he gave no order to discontinue use of the sheets and assumed they were being used.

Communication Problems

There were several serious violations of established communication procedures that resulted in excessive production lost time. These are:

1. Failure of the operator or control room to report bearing overheat to the maintenance foreman responsible for oiling. This might have prevented the failure.
2. The night shift foreman used poor judgement in not assessing this incident as a critical problem. He assumed that since another fuel pump was on the line that the problem could wait until the morning. The control room violated existing procedures in not notifying the Plant Manager of a reduction in tower output.
3. Concerning the location and delivery of replacement bearings - Subsequent investigation revealed that suitable bearings were in stock listed under a different equipment item. The stores supervisor stated that he would have picked this up immediately if he were notified of the existence of a critical problem, which he wasn't.

   In addition there was no reasonable excuse for the ordering of bearings being delayed until 3 P.M.

   A further foul-up caused by lack of communication between maintenance and stores resulted in the delivered bearings sitting in receiving for one hour before notification to maintenance.

The problems with communication were attributed to:

- Violation of existing procedures
- Poor judgement and lack of sensitivity to the criticality of the problem on the part of the several persons involved.

Recommendations

1. That the auxiliary fuel pump be replaced with a full sized unit, electrically interlocked.
2. That the oilers check-out procedure be reactivated after suitable reindoctrination of affected personnel.
3. That the communication problems of this incident be reviewed with affected personnel together with the existing critical problem reporting procedure.
4. That maintenance engineering look into the stocking problems surfaced by this incident.
5. That the advisability of disciplinary action be investigated further.
6. That modification procedures and failure feedback reporting procedures be reviewed with maintenance foremen because of the lack of records on an apparent previous bearing problem.

COMMENTS ON THE EXAMPLE

What frequently happens on critical problems is that the recording and analysis ends at step 3. Or even worse, the details of steps 1 through 3 are left out. The foreman in this case says "We fixed it; the incident is resolved." For example, the report might have been made verbally as follows:

1. What Happened?

The south main fuel pump broke down.

2. What Corrective Action Was Taken?

A bearing burned out tearing up shaft, flexible coupling and impeller. All parts were fixed and reinstalled by 12 noon on November 13th.

3. What Was Impact On Production?

39 hours downtime @ $5,000 per hour × .3 = $58,500.

The point to be made is that when a serious event like this occurs, we are not only interested in eliminating future reoccurrences, but we are also interested in the details of *why* the downtime lasted for 39 hours and how we can cut the repair time to a bare minimum if the failure reoccurs.

Unfortunately, the shortened story told in the above example is frequently enough to satisfy some people. This story says *nothing* about the thousands of dollars wasted by a part sitting in receiving, or of a supervisor arbitrarily deciding to delay the rebuild, or in the foul-up on preventive maintenance instructions. These problems are only surfaced and resolved by the inquisitive mind and a thorough investigation.

RELATED CASE STUDIES

CASE STUDY 2C: Process Plant Bearing Problem

Critical Problem Report

No. 513–013–01 Date 1/24/80 By F. Herbaty

1.0 What Happened?

Production worker noticed southeast bend pulley bearing smoking on belt #3 at 7 P.M. on January 21, 1980. Bearing was immediately greased. By 10 P.M., it became apparent that the bearing was burning up and could not be saved.

2.0 What Corrective Action Was Taken?

At the change of shift at 11 P.M., 2 men were assigned to replace the bearing. A Rex P-2215 bearing available in stock was installed. Belt #3 was shut down from 11 P.M. to 3 A.M. on January 22nd to accomplish this task.

The work was covered on W.O. #84866. The cause of failure noted on this work order was "lack of lubrication".

3.0 What Was the Impact on Production?

Belt #3 was down from 11 P.M. to 3 A.M., or four hours, for repairs. Since the silos were very low, this caused an almost immediate shutdown to both slurry mills, which were down until approximately 6:30 A.M. on January 22, 1974, or approximately seven hours. (One mill was down 4½ hours, the other 7 hours).

This outage did not immediately affect kiln production, but according to the superintendent, the decrease in slurry output *may* cut into kiln production at a later date due to low slurry tanks and numerous problems in slurry mill feed systems including belt #7.

4.0 What Was the Cause of Failure?

4.1 General:

Investigation revealed numerous items which may have caused the failure. These items are discussed in the following paragraphs.

4.2 History:

Information in the work order files is not completely clear, however, it appears that there has been approximately eight bearing failures at the counterweight bend pulleys in the past fourteen months. 2000 lbs. were added to the counterweight in 1976 and an unmeasured quantity approximately ten months ago.

According to the original design drawing (Hewitt Robins Sketch No. 1629), the original bearing specified was

"Dodge Type E Pillow Block Bearing"

The stock card for this bearing specifies a Rex P2215 as an equal to the original bearing. The original bearing, however, hasn't been ordered or stocked for the past seven years. Furthermore, the P2215 bearing is *not* equal to the originally specified bearing.

4.3 Bearing Seal:

On page 400 of the Rex Bearing catalog, different seals are rated as follows:

4—Excellent

3—Good

2—Fair

1—Limited

0—Not Acceptable

The existing situation at the bearings in question consists of an atmosphere of abrasive materials, frequently wet, causing a coating and buildup on the shaft. As a matter of fact, it is reported that these bearings are frequently submerged and running in this abrasive material.

The seal used on the Rex-P2215 bearing is a P seal which is practically the worst for the application having a 1 rating. The M seal has a 4 rating and is the one that should be used. This error is considered to be the major reason for failure.

4.4 Bearing Load:

The counterweight load is the principal load on the bearing in question. Its exact weight is difficult to find, however, the following calculation is an approximation:

	Pounds
Original weight from equipment file in engineering	7,240
Counterweight added in 1976	2,000
Counterweight added in 1978 (estimated at 20%)	2,000
Estimated build-up weight (from engineering files)	2,800
Subtotal	14,040
Add 50% for heavy service (see page 10-44 of Dodge Selection Guide Book)	7,020
	21,060

$$\text{Load per bearing (4 bearings)} = \frac{21{,}060}{4} = 5{,}265$$

$$\text{Vector component load } 1.5 \times 5{,}265 = 7{,}898$$

The allowable load of the bearing in use (Rex P-2215) is 7,175 lbs. In other words, it is too light.

It is important to note that the maintenance foreman over belt #3 states that the last time the bearing failed, the counterweight was *completely submerged* in stone and that it was cocked sideways a couple of feet thus binding it in its rails! This condition would place practically an infinite load on the bearing.

4.5 Bend Pulley Scrapers:

Presently there are scrapers on both bend pulleys which cascade product directly on top of the counterweight. In addition, there is spillage directly from the belt which dumps product on top of the counterweight and the bearings.

The original design attempts to take care of this situation by having a hole directly below the counterweight which leads into a chute which discharges the material onto belt #7. This design is unsatisfactory for the following reasons:

- It does not prevent buildup on top of the counterweight.
- The hole below the counterweight frequently plugs up.
- The chute leading to belt #7 is unsatisfactory since it allows spillage to accumulate on both sides of belt #7.

4.6 Bearing Misalignment:

According to the maintenance foreman, it is not unusual to cock the shaft as necessary to properly center the belt when reinstalling a bearing. In other words, there is need for a bearing with sizeable misalignment capabilities such as is inherent in the Rex type design.

4.7 Lubrication:

The work order states that the bearing failed due to lack of lubrication. A check with the oiling records shows that the "Discharge End of Belt #3" was lubricated on January 21, 1974, the date of the incident. A further check shows the same notation a week previous.

Questioning of the oiler involved confirms that the quoted statement *includes* the bearing that failed and that he did in fact oil the bearing on that date.

Presently, there is *no* written check sheet ordering the lubrication of the subject bearing. Instead, there is a verbal agreement to lubricate it once per week. This agreement is unsatisfactory since it is also considered proper to drop the lubrication if more pressing work presents itself. Basically, there is *no way* to satisfactorily audit compliance to this type of instruction or to confirm that the oiler is telling the truth.

The existing lube frequency of once per week appears to be too frequent and is probably required because of the incorrectly selected seal. When a bearing seal of the proper type is installed it is recommended that the lube frequency be reduced to once per quarter.

5.0 What Should Be Done to Prevent Recurrence?

5.1 Bearing Design:

The excessive bearing failures can be eliminated by proper bearing selection as follows:

- Radial Load Capacity - 10,550 lbs. at 100 RPM. See Rex Catalog, Page 399, for 9000 Series or Equal.
- Seal Type M or equal. See Rex Catalog, Page 400.
- Fully Self-Aligning Bearing required similar to Rex Design.

It is recommended that Rex MPA-215F (M Seal) be used.

It is furthermore recommended that a 2′ × 1′ hole be cut into the grating directly below the bearings to prevent buildup on the bearings. This hole can be partially closed by installing ⅛" × 1" bars on 3" spacing for safety purposes. Consideration may be given to installing a sheet metal cover over the bearing in addition to protect it from buildup.

5.2 Lubrication:

Initiation of the preventive maintenance program specified in the Manual will provide for specific "engineered" instructions *and* a basis for auditing compliance with the instructions.

5.3 Similar Bearings:

Engineering should make a study to determine which other bearings in the plant fall under the same circumstance as this bearing. These bearings should then be programmed for similar corrective action.

5.4 Product Spillage:

Product Spillage problems surfaced in this report should be made the subject of a separate study by engineering.

5.5 Analysis of Critical Problem:

The problem with the subject bearing is a serious one of relatively long standing. The fact that an investigation has not taken place automatically some time ago is a source of concern.

It is recommended that all production, engineering and maintenance personnel be indoctrinated in the purpose and details of the procedure for analyzing critical problems.

5.6 Bearing Replacement Method:

The bearing replacement time for this incident was four hours. Some foremen say that the job can be done in one and one-half hours. Since downtime is of major concern on this belt, it is recommended that a job method study be made to determine if there is any way to cut down on the replacement time.

The bearing in this case was burned out with a torch. This gives rise to the question of whether or not the shaft has been damaged to the extent that it is causing bearing failures.

5.7 Head Pulley Slippage:

It has been reported that despite the addition of counterweights two times that slippage is still experienced at the head pulley. This is worthy of a separate investigation.

5.8 Stocking Problems:

Someone, seven years ago, specified an "or equal" bearing that was not equal to the original design on several counts. Of course, the impact of this error is less because neither bearing is suitable for the job.

This occurrence should be used to illustrate to people making the "or equal" decision to be careful in the process and to make the necessary evaluation as a prerequisite to selecting an "or equal" item.

5.9 Counterweight Buildup:

Production should be instructed never to let counterweight buildup exceed two feet.

CASE STUDY 2D: Process Plant Critical Conveyor Problem

The problem illustrated by this example was not a single incident but a recurring problem and, therefore, does not follow the format previously described. In this case, management decided that the recurring problem was serious enough to warrant a special study.

1. General

Belt 7 is a critical belt since all product flows through the belt and there is no redundant belt.

An analysis of work orders and the log book for belt 7 reveals the following highly repetitive problems:

a. Repairs to belt - tears and clips and saddles.
b. Buildup at tailpulley.
c. Buildup at headpulley.
d. Renewal of return rollers.
e. Cleaning of dirt from rollers.
f. Bearing failures
g. Dynamatic problems
h. Coupling problems

2. Product Spillage

One of the biggest causes of downtime results from product spillage at the tail end. Product is spilling out of worn skirts and holes in chutes. It is believed that this spillage is the primary cause for problems a, b, d, e and f.

It is firmly believed that performance of the tasks described in the attached preventive maintenance instruction will virtually eliminate these problems.

In order for this preventive maintenance instruction to be effective, belt 7 must be put into proper condition by doing the following:

a. Renew skirts
b. Reline chutes on brushes and wobbler discharges.
c. Countersink clips and eventually replace with vulcanized splices and repairs.
d. Reinstall scrapers
e. Close opening in table top above counterweight.
f. After all of the above items have been taken care of; the tail end area should be thoroughly cleaned.

A special problem exists with product spillage from belt 3 to belt 7. It is recommended that this be the subject of a separate study by engineering.

Another major spillage problem occurs at the head end. For the period April 11, 1973 to January 29, 1974, there were forty three downtime occurrences to dig out product accumulation under the return strand at the head end. This problem is caused by material being carried back on the return strand. The problem is aggravated by a relatively small clearance between the return strand and the floor plate. There are several possible ways to eliminate this problem:

a. Countersink belt clips and/or install vulcanized splices and repairs. This will permit installation of belt scrapers and/or brushes.

b. Modify the shape of the floor plate at the accumulation to prevent buildup.

It is recommended that engineering study this problem and come up with the most advantageous design.

3. Bearings

There have been nine (9) pulley bearing failures in the past fourteen (14) months. The original design on these bearings is as follows:

"Rex ZPA-315-F"

This bearing is satisfactory from load and misalignment standpoints but has an unsatisfactory seal. The seal for this very severe abrasive dust condition should be a type M seal.

Approximately two (2) years ago, the purchase of this bearing was discontinued and the following substituted:

"Rex P-2315F"

This bearing is even more unsatisfactory since its allowable load is only one-half of the original design.

Furthermore, it was observed that some of the installed bearings were of the 2-hole type instead of the original 4-hole type, and were mounted on unsatisfactorily designed bases.

It was noticed that some of the 4-hole installed bearings were mounted on 2-hole bases with inadequate base support area.

In addition, several of the bearings were submerged in product spillage to the point where the shaft and the bearing were being worn.

Also, connections were made to grease points on bearings located on the opposite side of the catwalk with inadequate rubber hose which was fastened to pipe nipples with twisted wire.

It was also observed that there is a high rate of bearing failures on the return and troughing idlers. It is believed that this is caused by an insufficient frequency of bearing lubrication (this frequency was increased from yearly to semiyearly in the attached preventive maintenance instruction.)

It is recommended that engineering make a complete study of *all* bearings on belt 7 and come up with the proper bearing design. The study should include the following:

- Load calculations and requirements
- Seal specs (M type recommended)
- Misalignment specs (Fully misaligned capability type recommended)
- Bearing base design (Adjustable type recommended)
- Proper design of lube point extensions

In order to alleviate the problem of spillage accumulation around bearings, it is recommended that:

- A portion of decking be removed around bearing base so that accumulation can fall through.
- That a metal shield be installed over bearings.

4. Dynamatic

Four (4) failures of the dynamatic have been experienced in the past fourteen (14) months. This has been a consistent problem in the past years.

The impact of a dynamatic failure is very severe because, when the dynamatic fails (since there is no spare), it is replaced by a solid jack shaft. This causes approximately four (4) hours of unscheduled downtime. The result of this change is that there is an instantaneous shock load on the motor, couplings, gear box and the entire conveyor systems, which has caused:

- Coupling failures between motor and gear box
- Motor failure and motor shaft twisting
- Belt splice damage
- Excessive belt strain
- Causes top strand to rise off idlers during startup, hitting magnet

An investigation reveals that all dynamatic failures have been in the bearings. Within the past month, a redesigned dynamatic has been installed. The redesign includes the addition of two (2) extra bearings. It is hoped that this modification will improve the situation. An investigation of the possible causes of bearing failure reveals the following:

a. According to the Dynamatic Company, improper alignment procedures are a major cause of bearing failures. It is, therefore, recommended that correct alignment be accomplished with the use of dial indicators.

b. It has been reported by the maintenance foreman that the internal components of the dynamatic have been so heavily encrusted by dirt that it takes two (2) days to scrape the parts clean. This situation very likely causes an unbalanced condition and vibration and contributes to bearing failures. In order to alleviate this condition, it is recommended that a duct be installed through the nearby roof, complete with filters and a gooseneck to lead clean air into the dynamatic air intake, as opposed to the present air which is saturated with erosive rock dust.

c. An investigation of the causes of bearing failure reveals that the present type of grease has a tendency to channel away from the bearing races. Mobil Oil Company has recommended that a new and improved product be used. This is Mobile Grease #27. This product is currently being tested in a smaller dynamatic.

d. It has been reported on numerous occasions that the motor and dynamatic base bolts have loosened allowing misalignment. In the past when this has occurred, the bolts have been tightened without going through realignment procedure. It is recommended that realignment be accomplished whenever this occurs and that the bolts be retightened and torqued to their maximum specifications and that the motor, dynamatic and gear box be doweled to base.

In addition, a weekly inspection has been specified of those bolts in the attached preventive maintenance instruction. (See Figure 2-12)

Belt 7: Summary of Recommendations

Item No.	Description	Responsibility
1	Implement PMI	Mechanic maintenance
2	Replace chute liners	Mechanic maintenance
3	Countersink clips and reinstall scrapers	Mechanic maintenance
4	Close opening in table top over counterweight	Mechanic maintenance
5	Thoroughly clean tail end area	Operational maintenance
6	Investigate spillage from Belt 3 to 7 and recommend design change	Engineering
7	Investigate spillage at headend and recommend design change	Engineering
8	Investigate bearing design, recommend new design, implement new design, correct stock problems	Engineering
9	Follow-up on success of bearing redesign on dynamatic	Mechanic maintenance
10	Use correct realignment procedures on dynamatic and motor (check present alignment and dowel to base)	Mechanic maintenance
11	Install dynamatic clean air duct	Mechanic maintenance
12	Implement new dynamatic grease	Mechanic maintenance

CASE STUDY 2E: The Handrail Welders

This story took place in a process plant. There was a line of huge smelters, many stories high on one wall of the smelter building and a line of Bessemer Converters on the opposite wall. The two groups of equipment were separated by a narrow crane aisle.

The traveling crane, which transversed the length of the building, was located at the underside of the roof about seven stories high. The crane carried a 20 ton pot, used to transfer molten ore from the smelter to the Bessemer Converters for further refining of the ore.

Sheet 1 of 1

330207 MACHINE NUMBER	P. M. INSTRUCTION	WORK ORDER NO.
MACHINE DESCRIPTION Belt 7		COST CENTER NO.330

		MECHANICAL					ELECTRICAL					LUBRICATION				
		W	M	Q	S	A	W	M	Q	S	A	W	M	Q	S	A
1.	Replace all skirts on tail end (10")			8												
2.	Adjust Skirts and Scrapers	2														
3.	Inspect Crusher Stone Box and Chute Liners. Replace as Necessary.		¼													
4.	Starting at Head Pulley and Proceeding to Tail Pulley, Inspect the Following: • Check Tightness if Motor, Dynamatic and Reducer Base Bolts for Tightness. • Check Head Tail, Counterweight, and Bend Pulley Bearings. • Check all Top and Return Idlers • Check for Excessive Buildup of Any Type	1														
5.	Check Belt Running While Empty	½														
6.	Check Safety Pull Switches			½												
7.	Lubricate Motor Bearings								½							
8.	Inspect and Clean Motor Starter										2					
9.	Lubricate Pulley Bearings											1				
10.	Lubricate Dynamatic													½		
11	Lubricate Troughing and Return Idlers														32	
	TOTALS	3½	¼	8½					½		2	1		½	32	

Figure 2-12: PM instruction.

While studying the maintenance organization chart, it was discovered that there was a group of fifteen (15) welders whose sole job was to repair handrails on platforms and structural steel members of the smelters and converters which were being smashed by the ore pot hanging from the crane as it worked the aisle.

When asked why nobody stopped this seemingly unnecessary destruction of facilities. The plant personnel looked upon the author as a wise guy and invited me to solve the problem.

I had just talked the president into starting the Analysis of Critical Problem procedure and was anxious to find a good example for the first analysis.

I was shocked to discover that the fifteen (15) welders had been performing this work for ten (10) years. I needn't have been shocked because over the course of several years, I was to discover that many plants suffered serious problems that were reoccurring over many years. The personnel had come to look upon the critical problem as just part of the job of running the plant, in a manner similar to the need to periodically change spark plugs.

After arranging for a meeting of all top management and explaining that having fifteen (15) welders repairing smashed handrails for ten (10) years was not like changing spark plugs. It was a *major critical problem* that should be stopped immediately. I recounted my experience at other plants where critical problems were allowed to happen only once! I expressed confidence that the problem could be eliminated by myself with their help and cooperation. I immediately logged the problem.

I decided to visit and ride with the crane operator. I made it to the crane level at the underside of the roof and almost immediately fled in sheer panic unable to breathe. The underside of the roof was filled with poisonous fumes from the smelter and converters and the temperature must have been 110°F. I returned later with oxygen breathing apparatus. I could not understand how the crane operator could stand it. He said you get used to it and jobs were hard to find and that is why he stayed.

Investigation soon revealed that all twenty (20) roof ventilator motors were out of commission and were operating on natural draft only. The following was immediately recommended and approval to implement was obtained.

1. Repair all roof ventilator motors
2. Sealing of leaks on smelters and converters.
3. Enclosing and air conditioning of crane cab and providing electrostatic air cleaners.
4. Provide oxygen breathing apparatus in crane cab for emergencies.

5. Providing of windshield wipers and washers for crane cab operator so that he had a fighting chance of seeing what he was doing.

Meanwhile, I found out that many of the production foremen had added catwalks that stuck out in the crane aisle so they could see what was going on. All of these were removed after determining they were not necessary. The crane operator had a tough time dodging all these catwalks.

The handrail stanchions were redesigned, putting springs in the handrail sockets so that if the handrail was hit by the 20 ton crane pot, it merely bent over and returned after the crane had passed.

All of these action items reduced the problem to a minor manageable one.

CASE STUDY 2F: The Screw Conveyor Bearings

The author was asked to look into the failure of several hundred bearings per year on the screw conveyor receiving particulate matter from a battery of huge electrostatic precipitators in a large process plant. These failures had been occurring since the plant was started ten (10) years previously.

After interviewing the maintenance personnel familiar with the problem, it was discovered that the bearing was inside the screw conveyor itself and was, therefore, inundated in the material being conveyed and was subjected to a temperature of 350°F.

After deciding to spend a few hours looking into the file on the ESP's, I soon found a letter written by the chief engineer of the firm that designed the plant, (about ten years previously) to the company that was to provide the screw conveyor. The chief engineer gave the specifications on the bearings he planned to use on the screw conveyor. The return letter stated that the bearings anticipated for use had seals that were good for only 250°F and would fail under the design temperature of 350°F. Guess what? The process plant was using and stocking the bearings with 250°F seals. The design firm never followed up on the letter from the supplier and nobody in the process plant had enough intellectual curiosity to read the files where the answer to the problem was waiting for anyone interested enough to look.

CASE STUDY 2G: The Ore Belt Cutting

It was the first day of my assignment to make a maintenance study at a large process plant and was being conducted on a tour of the facilities. The guide was pointing out one of the very large ore carrying belts. There were six of these expensive belts all being fed by overhead hoppers by gravity.

As I watched, I noticed a large tear developing in the belt as if some giant knife were cutting it. I immediately showed this to my guide who pushed an emergency stop button.

The next day I came back to the site to find out what was causing the tear. I walked up to the level above and talked to the operator where the six hoppers were being fed by overhead cranes. The operator explained that it was his job to keep the hoppers flowing freely. This was a problem, since the ore would periodically cake-up and refuse to flow down by gravity. The tool he used to prevent this was a compressed air lance. It consisted of a 15 foot long, 1 inch diameter standard galvanized pipe with shut-off valve fed with 100 PSI compressed air through a long hose. The hose permitted him to service all six hoppers. When one of the hoppers started to clog up, he stuck the pipe into the inverse pyramid shaped hopper, turned on the compressed air and wiggled it around until the ore ran freely again. Asked if they had found the reason for the belt tear from the previous day, the operator said that the end of his pipe lance had broken off and flowed down into the horizontal belt where it jammed against the sides of the hopper. The broken end of the pipe then acted like a knife, cutting the belt. When asked how often this happened, he said it happened about twice per week and had been going on for seven years.

If I hadn't seen many things like this before, I would have been shocked. That day I recommended that vibrators be installed on the sides of the hopper. This solved the problem.

CASE STUDY 2H: The 1000 HP Motor Fire

As usual when starting the day, I checked with the Chief Planner Scheduler to see if anything unusual had happened during the night. He said yes, a 1000 HP sychronous motor had caught on fire and was completely destroyed. A spare motor had been installed and was already in operation.

I checked with the maintenance foreman to find out what the cause of the fire was. He said they were unable to determine the cause and that men were now examining the burned out motor.

I immediately called the General Manager and recommended that the motor be shut down, since the cause of the fire had not been determined, and there was a possibility that the new motor would burn up for the same reason. He agreed and had the motor shut down.

It was later determined that the fire was caused by a faulty protection device. This device was repaired and the motor was then returned to operation.

Examples 2B through 2E illustrate the consequence of not investigating a serious incident or of an inadequate investigation. It also demonstrates that it is not rare for personnel to consider even major failures as commonplace events that "one must expect in our business" and feel no obligation to determine the cause and eliminate recurrences.

The most important point of this section is to differentiate between run of the mill failures which should be handled as described in Chapter 10, Failure Analysis, and major failures or catastrophes, which must be investigated immediately as described in Chapter 11, Analysis of Critical Problems, to insure that they do not reoccur.

12

The Principle of Simultaneous Downtime

The Principle of Simultaneous Downtime (POSD), a concept conceived by the author, is particularly important in maintaining large, complex, expensive equipment items with high downtime costs.

The principle is quite simple—component failures should be synchronized to occur simultaneously, thereby limiting breakdown to only one occurrence instead of several.

In the ideal case, part A and part B each have a failure period of exactly 6 months. Downtime caused by failure of each part lasts 8 hours. Annual downtime caused by the two parts failing at different times will be: 2 parts × 8 hours per failure × 2 failures per year = 32 hours.

Replacing both parts when one fails will synchronize the failures so they wil occur simultaneously. Annual downtime on the equipment can then be reduced from 32 to 16 hours.

However, in practice the concept is more complicated.

For example, consider the case of a 30-foot-diameter, 300-feet-long rotary kiln, installed in a process plant running 24 hours a day, 7 days a week. The material produced is in short supply and all product can be sold. Downtime on the kiln, therefore, causes an irretrievable loss in sales.

Downtime usually lasts two weeks, and the downtime cost is $10,000 an hour. Total cost for each downtime occurrence is: $\$10,000 \times 24 \times 14 = \$3,360,000$.

The loss in sales dollars when the kiln breaks down is tremendous.

A thorough failure-recording system will pinpoint which parts have failed in the past and how much downtime the failures have caused.

The maintenance manager must rely on experience in choosing the best maintenance approach to increase kiln productivity and reduce downtime.

In this case, he recognized that brick repair was the major cause of downtime and had replaced the original brick and insulation with different types during repair.

He was also familiar with the many problems common to the combustion system, the precipitators, and the input/output mechanical systems.

His first approach at reducing downtime was:

- Use as many repairmen as necessary (the entire maintenance department and some production people, if required) to repair the kiln as fast as possible
- Maintain an adequate stock of spares
- Make only those repairs necessary to return the kiln to operation as soon as possible.

With this approach, kiln productivity was 70 percent. The kiln was breaking down frequently for various reasons. Although the average length of the downtime period was reasonably small, the number of occurrences was high. And, the plant manager was unhappy because kiln productivity percentages in several other plants were higher than 70 percent.

The maintenance manager chose another approach. He noticed that brick failure would occur adjacent to repairs of previous failures. Repairmen were instructed to examine the condition of the brick adjacent to the area of failure and, if warranted by the inspection, renew an entire band of brick, perhaps 5, 10, or 15 feet wide, instead of replacing only the failed brick. Records were kept of which areas were renewed.

A mechanical maintenance man, repairing a failed bearing on a primary air fan for the kiln's main burner, suggested to the maintenance manager that the bearing could easily have been replaced during the last downtime occurrence. The maintenance manager then ordered that whenever the kiln was down for repairs, a team of experts would thoroughly inspect the brick, the insulation, and the mechanical, electrical, electronic, and structural systems to determine if any components were likely to fail in the near future. The team would make recommendations for replacing or overhauling weak parts, if the repairs could be made within the time required to repair the part causing the downtime. And, the economics of replacing or repairing parts with longer repair times would be evaluated and appropriate orders issued.

Within 6 months, this revised approach increased kiln productivity from 70 percent to 80 percent.

At this point, the maintenance manager was concerned with further increasing kiln productivity. In annual dollars the remaining 20 percent represented approximately: 0.20 × 8760 hours per year × $10,000 lost production cost per hour = $17,520,000. Some of this money could be saved by using POSD.

The basic procedure for applying the principle of simultaneous downtime is:

1. Breakdown repair orders should be sorted according to failure items.
2. Records of each type of failure should be arranged chronologically.
3. Times between failures for each type of occurrence should be calculated.
4. Average times, and the longest and shortest times between failures should be calculated.

5. Results should be tabulated (see Figure 2-13).
6. Data should be analyzed so decisions about synchronizing failures can be made.

The accompanying table reveals failures of kiln components 2, 3, 4, 5, 6, 7, 8 and 11 should be synchronized because:

- Variations of the high, average and low times between failures are very small
- Ratios of downtime costs to repair costs are very high, ranging from 53.3 to 6.7. For example, cost of downtime caused by item 8 is 53.3 times that of the actual repair.

Item 1 was excluded from further consideration for two reasons:

1. Time between failures was too variable.
2. Ratio of downtime to repair costs was too low and the repair cost was very high. This brick area should be replaced only in the event of a failure.

The decision to synchronize failures is difficult because of the variations in time between failures of many components. Careful study of the time between failures, downtime cost, repair cost, and the ratio between these costs is necessary.

Another consideration is whether a component failure will cause catastrophic failures of other parts and create safety hazards. If it will, the repair should be made during a planned shutdown. The repair should be timed to occur before the mean time between failures, and instrumentation should be installed to give advance warning of failure.

Application of the POSD approach in this case would have increased kiln productivity well above the 80 percent mark. A sound preventive maintenance program should be in effect for all kiln auxiliary or redundant equipment items that can be shut down for preventive maintenance without shutting down the kiln.

No.	Component	Time Between Failures Months			Downtime for Repair, Hours	Cost X: Average Downtime Cost, Dollars	Cost Y: Average Repair Cost, Dollars	Ratio: Cost X/ Cost Y
		High	Average	Low				
1	Brick Area A	10	4	2	336	3,360,000	1,000,000	3.4
2	Brick Area B	4	4	4	280	2,800,000	400,000	7.0
3	Brick Area C	4	4	4	200	2,000,000	300,000	6.7
4	Insulation Area A	4	4	4	150	1,500,000	100,000	15.0
5	Insulation Area B	4.1	4	4	125	1,250,000	70,000	17.9
6	Insulation Area C	4	4	4	80	800,000	60,000	13.3
7	Burner Part A	4.2	4.1	4.1	16	160,000	4,000	40
8	Burner Part B	4	4	4	8	80,000	1,500	53.3
9	Burner Part C	8	6	5	4	40,000	10,000	4.0
10	Cooler Part A	12	7	3	40	400,000	40,000	10.0
11	Cooler Part B	4.5	4.4	4.3	30	300,000	10,000	30.0
12	Cooler Part C	17	11	6	10	100,000	8,000	12.5

Note: The above data is purely hypothetical and not related to actual numbers.

Figure 2-13: Principle of simultaneous downtime chart.

POSD could also be used as the basis for scheduling planned periodic shutdowns for repairs. The frequency and duration of downtime, as well as the items to be repaired, can be determined by POSD.

A variation on the concept of periodic overhauls is the annual total plant shutdown, during which *everything* that needs it, is overhauled or repaired. This approach violates the POSD, since components are overhauled regardless of time between failures. This approach, however, may be justified in some cases where there is a seasonal drop in demand for the plant product close to zero demand. Another reason might be that the raw material feeding the plant is only available during a part of the year.

13

Preventive Maintenance (PM)

GENERAL

Today's technology is advancing at an incredible rate. Plants are becoming ever more automated, resulting in an increase in the ratio of maintenance to production workers and job complexity. In the plant of tomorrow, the majority of workers will be those highly skilled in maintenance. Therefore, careful study to improve methods and reduce maintenance costs is imperative. Achieving these objectives will help save billions of maintenance dollars.

Specifically, such a study should examine methods used by industry to write preventive maintenance instructions. Although much has been published about selling, managing, scheduling, setting up paperwork systems, etc., little has been written about the most important part of the preventive maintenance program — the quality of the preventive maintenance instruction itself.

Finding personnel with all of the qualifications necessary to the writing of good preventive maintenance instructions is always a problem, and the smaller the number of people authorized to do so, the more the problem is compounded since in that case, each preventive maintenance writer must cover a broader area. If a shortage exists, it should be mitigated by outside or in plant training. In any event, management should put its most qualified people on the task of writing preventive maintenance instructions.

One of the most widely used reference sources for writing preventive maintenance instructions is the manufacturer's maintenance instruction book.

It is important when consulting this manual that the manufacturer's philosophy concerning maintenance costs be understood. First of all, the manufacturer designs a machine to be as reliable and maintenance free as manufacturing cost and competition will permit. Once the design is fixed, he may then tend to overwrite preventive maintenance instructions to, as nearly as possible, approach the zero breakdown level. He does this in order to achieve a reputation of reliability for his machine. He may be more concerned about the machine's reputation than he is about costs. Of course the owner is interested in both. Furthermore, it would be difficult for the manufacturer to cover all the variable conditions his machine may be subjected to in one set of instructions.

As a general rule, attempting to write preventive maintenance instructions to achieve zero breakdowns is as impractical and uneconomical as it is to attempt to achieve a zero reject rate in a manufacturing operation. Consider the example of the automobile manufacturer who recommends that the oil be changed every 2,000 miles. What would be the change in the car's total maintenance cost and reliability if oil changes were recommended every 4,000, 6,000, 8,000 or even 12,000 miles? Negligible, most likely, if the owner replaced the car every two years.

The difference between changing the oil every 2,000 miles versus every 12,000 miles represents a cost multiplier of six. This single cost may seem insignificant but, if applied to a plant with 100,000 units of equipment on which excessive preventive maintenance instructions are written, the total extra maintenance cost could be in the millions of dollars.

What can be done to reduce these total costs? Let's focus on a single piece of equipment and plot curves to keep track of the variation in maintenance costs and downtime percentage to supplement an analysis of breakdowns and the writing of maintenance instructions to prevent these breakdowns. Figure 2-14 shows three curves plotted against preventive mainte-

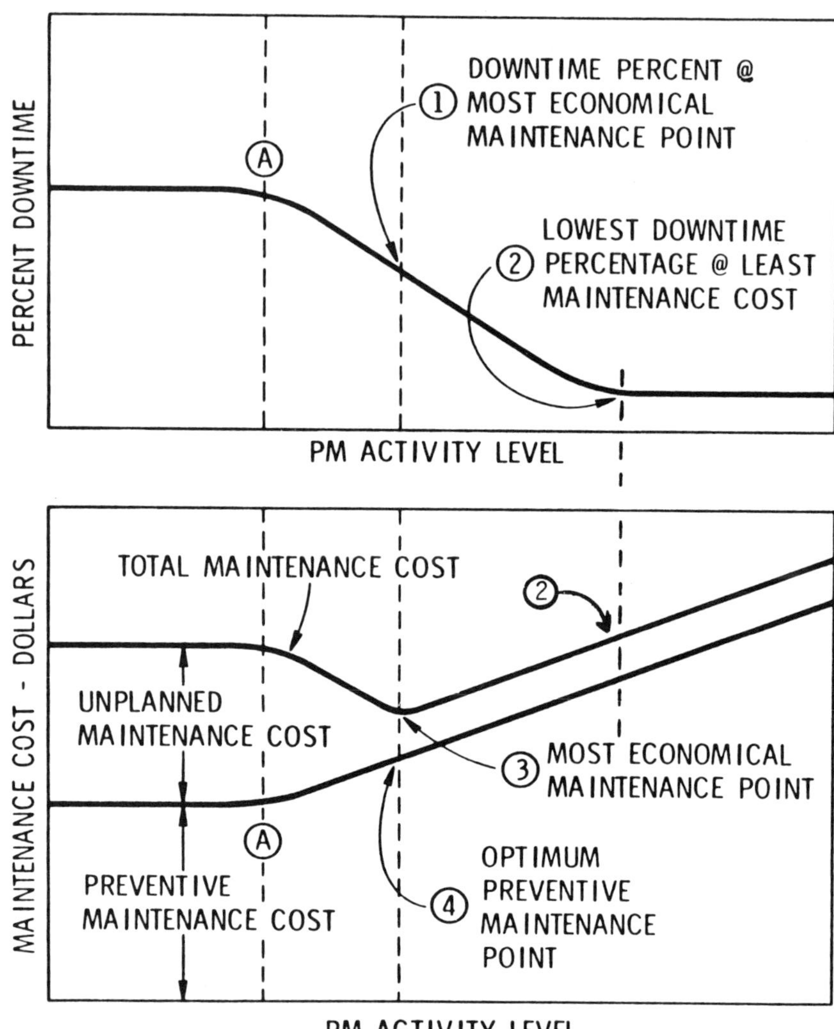

Figure 2-14: PM optimization.

nance activity level, i.e.: downtime percentage, preventive maintenance cost and unplanned maintenance cost.

How can the chart be interpreted? First, some definitions are in order. Preventive maintenance cost is the cost of maintenance resulting from preplanned, written maintenance instructions. Unplanned maintenance costs are all other related costs and are evidence of the failure of the written preventive maintenance instruction. Downtime percentage is the percentage of available machine time that the machine is broken down and unavailable for use. Construction, operation and modification costs are excluded from consideration.

Reading from left to right on all three curves up to Point A represents the established maintenance system, and includes both preventive maintenance and unplanned maintenance costs with a certain percentage of downtime. An analysis of the breakdowns in the historical record file maintained on each piece of equipment serves as a basis for improving, adding, or even reducing preventive maintenance instructions.

For example, using the automobile analogy again, if spark plugs fail frequently with consequent downtime, it would be logical to add a written preventive maintenance instruction to replace the spark plugs at periodic intervals to prevent breakdown. Point A represents the point at which additional preventive maintenance instructions are included in the maintenance program. If the original preventive maintenance program were modest in scope, and if the added instructions were intelligently written, several trends should become evident. One is that preventive maintenance costs will rise in direct proportion to the cost of implementing the additional instructions. However, the unplanned maintenance cost and the downtime percentage should drop, as well as the total maintenance cost. The reason is that planned preventive maintenance is more efficient. An individual assigned to the work is given all the instructions, materials, tools, etc., required to do it quickly. In the case of unplanned maintenance, the worker must take time to identify the problem and diagnose the fault before he can begin repairs. Frequently, he must return to the shop to get the needed material and tools which may not be in stock, further increasing downtime and costs.

Further, unplanned breakdowns often pyramid into additional breakdowns with resulting higher cost. For example, an unplanned failure of a motor bearing may cause the shaft to drop, and seizing of the rotating part. The result might be a cost rise from a $10 bearing replacement charge to a complete rebuilding cost of thousands of dollars. If preventive maintenance instructions are continually added and frequencies increased, the point of saturation is eventually reached and the preventive maintenance task is no longer economical (see Point 3). At this time a reversal in the downward trend of total maintenance cost will occur and costs will increase in direct proportion to the added preventive maintenance cost.

Point 3 is the most economical maintenance point; Point 4 is the optimum preventive maintenance point. Point 1 is the downtime percentage produced at the most economical point. Determination of the points at which economy is maximized is the purpose of the study and can result only from an intelligent analysis of maintenance records, coupled with an analytical approach to varying the preventive maintenance effort.

In some cases where critical equipment is involved, maintenance cost is of secondary importance, the lowest possible downtime being the primary consideration because it affects production loss. Point 2 represents the lowest possible total maintenance cost at which the lowest downtime can be achieved. In other words, increased preventive maintenance effort beyond this point produces no decrease in downtime. As an aid in determining the level of maintenance, it is helpful to divide facilities and equipment into three categories:

> *Category A*—Items of very small dollar value where no preventive maintenance program is planned. A $10 electric fan is an example. Here, rather than perform preventive maintenance, it is more economical to let the fan run until breakdown and then either replace or repair it . There is no preventive maintenance program for this category, only breakdown maintenance.
>
> *Category B*—Equipment of relatively large dollar value where a reasonably low downtime is required and maintenance of the equipment at the most economical point is most important. This category has a normal preventive maintenance program.
>
> *Category C*—Equipment that must be operated at the lowest achievable downtime because of the critically of function or cost of production

downtime. This category must have a special preventive maintenance program, plus an aggressive maintenance optimization program.

With this approach, system and paperwork complexity are balanced against need. Where there is no need, the system and paperwork are at zero. Where there is a requirement for normal perservation of facilities and a reasonable downtime, a preventive maintenance program is installed. Where production lost time costs are of overriding importance, a maintenance optimization system is used.

Normally, zero downtime can be achieved only by placing redundant units side by side and so interlocking them that one unit starts up if the other fails. In this case, a decision must be made whether to maintain the equipment at its most economical point or minimum downtime point.

With the maintenance cost curves analyzed, effective preventive maintenance instruction can now be prepared. If a maintenance man who is familiar with a particular machine were asked to write a set of preventive maintenance instructions which might achieve Point 3 (lowest total maintenance cost) and yet another set to achieve Point 2 (lower downtime), he probably would be unable to do it. It would take a miracle to come up on the first try with the optimum instructions to satisfy each objective. Although he might come close to it by chance, he would never know it if the cost and downtime data were not plotted on graphs such as those in Figure 2-14.

Unfortunately, this is seldom done. Little cost analysis is applied to most plant maintenance programs to determine how economically efficient the individual machine maintenance program is. The dollars wasted through lack of such analysis can add up to significant costs.

Optimization of preventive maintenance, as represented by the finding of Points of economy 1-4, can be accomplished through experimentation and with the help of a well qualified maintenance man who is familiar with the particular machine under study. It must be done in the following manner: Starting at Point A, the historical record file should be analyzed and preventive maintenance instructions added to prevent future failures of the type noted on the record card. In addition, if a preventive maintenance instruction already covers the point of failure, but breakdowns still occur, the preventive maintenance frequency should be increased.

Conversely, if a preventive maintenance instruction covers an item or function on which there has been no breakdown for a long period, either the preventive maintenance frequency should be reduced or the task eliminated. By varying preventive maintenance frequencies and tasks, and by plotting the effect on total maintenance costs, you will be able to determine Point 3. The set of preventive maintenance instructions which are in use at Point 3 will be the instructions that produce the most economical total maintenance cost.

By continuing this variation and increasing instructions and frequency past Point 3, Point 2 will be reached. This represents the lowest maintenance cost at which the lowest downtime can be accomplished. The maintenance manager is thus able to maintain his equipment at either the most economical point or the point of least downtime with corresponding minimum cost based on the criticality of the equipment.

Mandatory vs. Non-Mandatory

Maintenance instructions can be divided into mandatory and non-mandatory types. Again, using the automobile analogy, spark plug maintenance can be considered to be an example of a nonmandatory type.

Consider the fact that an eight-cylinder car has eight spark plugs which are not likely to fail simultaneously. Actually, it is an infrequent occurrence when only one fails completely. Rather, they all begin deteriorating at the same time. If there were no specific instructions calling for periodic renewal of the spark plugs, the result would merely be a gradual decline of engine performance until it was finally found necessary to change the spark plugs. Because nothing serious will occur if you do not change the spark plugs on a prescribed or planned basis, the instruction is considered to be non-mandatory.

On the other hand, an example of a mandatory item would be the brakes on a car. If there were no preventive maintenance system on the brakes there would be no assurance that brake failure would not occur without warning. Therefore, it is mandatory that you write a planned maintenance instruction calling for periodic brake inspection and an all-out overhaul of the system after so many miles of operation. Understanding of these categorizations and an intelligent appraisal of historical feedback information will help the preventive maintenance engineer make proper decisions concerning preventive maintenance coverage.

Pyramiding vs. Nonpyramiding

Another type of maintenance instruction is known as pyramiding and nonpyramiding. Consider the following example. A fuse blowing in an electric circuit is an example of a nonpyramiding type of breakdown. When a fuse blows, that is the end of it—nothing else happens. An example of pyramiding type of failure might be the previously cited failure of a bearing—the main bearing in a engine, for instance. If it fails, the shaft may drop and the entire engine may seize, thereby causing failure and increased cost. Understanding pyramiding will aid the preventive maintenance engineer in making decisions concerning relative importance of maintenance tasks.

Inspection vs. Positive Instructions

There is much controversy concerning simple inspection steps versus "positive" maintenance instructions. Many feel that a preventive maintenance program should consist of inspection items only and very minor maintenance work. Others feel that preventive maintenance should consist only of "positive" tasks.

To illustrate the difference between the two, consider the maintenance of a belt drive. If planned maintenance were handled by inspection only, the task would call for inspection of the belt drive at a certain frequency. The instruction would probably say "Inspect the belt to see if it is in good condition and has the proper tension. If it is in bad condition, replace it." On the specific "positive" task basis, the instruction would merely call for replacement of the belt.

Again, an understanding of these two categories will enable the preventive maintenance engineer to make an intelligent appraisal of the maintenance needed. The belief that preventive maintenance must consist of either 100 per cent positive tasks or 100 percent inspection tasks is unsound. Each maintenance requirement should be considered individually. In certain cases, an inspection instruction is the correct approach; in others, a positive instruction is preferable.

Returning to the belt drive example, and using the inspection approach, assume historical records show that the belts have been failing on the average of once per year. The inspection might be prescribed once per month and require round-trip travel time of 20 minutes, plus two minutes on the job. A yearly belt replacement cost of $60, in this instance, adds up to an annual total of $300 for inspection and belt replacement.

In the case of a positive instruction, belt replacement might be specified every nine months at a yearly cost of $90. In this case, it is apparent that the positive instruction is more economical.

The following paragraphs divide the task of writing preventive maintenance instructions into its several sub-elements for the purpose of making clearer the various processing steps and their logical sequencing.

PREVENTIVE MAINTENANCE SUPERVISOR (PMS)

It is essential that a single person be designated to manage the preventive maintenance program. This is a full time assignment which should be given to one of the best men in the department. The preventive maintenance supervisor is responsible for managing the implementation, operation and control of the preventive maintenance system which includes:

- Maintenance of the criticality coding list with the help of engineering
- Operation of the failure recording systems

- Development and maintenance of the failure coding system
- Preparation and follow-up of critical problem reports
- Preparation, checking, training of personnel and scheduling of the implementation of daily inspection route sheets
- Preparation, checking, training of personnel and scheduling of the implementation of preventive maintenance instructions
- Coordination with the computer department on problems related to the computerization of the preventive maintenance system
- Analysis and correction of maintenance problem areas
- Selection, training, scheduling, expediting and providing guidance to personnel involved with the preventive maintenance system
- Reporting on status of the preventive maintenance systems to the maintenance manager
- Coordination of preplanned outages
- Scheduling of analysis and scheduling work loads
- Conducting of simultaneous downtime studies.

IDENTIFY EQUIPMENT

The first step is to identify equipment to be preventive maintenanced and the priority of program implementation.

Categorization of equipment by criticality should be accomplished as explained in Chapter 5.

The most critical equipment items should be implemented first.

The preventive maintenance supervisor should then prioritize the items that are most critical to determine which to work on first.

ESTIMATE INSTRUCTION WRITING TIME

A rough estimate of the instruction writing time should be made for each of the prioritized items by the preventive maintenance supervisor for scheduling purposes.

ASSIGN RESPONSIBILITY

At a meeting chaired by the preventive maintenance supervisor with representatives of the maintenance, production and engineering departments, specific people should be assigned to write instructions for each of the equipment items.

The writing effort should involve people of the following types:

Maintenance foreman

Maintenance staff

Engineering.

SCHEDULE WRITING TASK

The preventive maintenance supervisor shall then prepare a schedule of the preventive maintenance instruction writing task utilizing the following information:

- Priority listing of the critical items
- Estimated hours per equipment items
- Part time capacity of assigned individuals.

The preventive maintenance supervisor will be responsible for:

- Maintaining a Gantt chart of progress
- Submitting weekly progress report to the maintenance manager

- Expediting task accomplishment
- Acting as a preventive maintenance consultant to assigned personnel
- Checking completed instructions.

IDENTIFY AREA RESPONSIBILITY

Maintenance responsibilities should be assigned and coded as follows:

Area Maintenance Responsibility Codes

Code	Area	Foreman
1	Maintenance Area 1	Jones
2	Maintenance Area 2	Smith
3	Maintenance Area 3	Smicek
4	Maintenance Area 4	Collins
5	Maintenance Area 5	Martello
6	Maintenance Area 6	Bauer
7	and so on	

These code numbers should be the numbers used for the first digit of the Work Order Number. The purpose of this is twofold:

- To assign specific maintenance responsibility by area
- To permit computerized segregation of preventive maintenance instruction and scheduling by area.

Specific names of foremen should be assigned to each of the areas. *They* shall be responsible for the maintenance program in their assigned areas. The preventive maintenance supervisor shall be construed as being staff support to these people.

ASSEMBLE REFERENCE MATERIAL

The writer of a preventive maintenance instruction for a specific item of equipment should systematically assemble reference material as follows:

- Manufacturers operating and maintenance manuals
- Recommended spare parts list
- Historical breakdown repair file
- Correspondence
- List and description of design changes
- Discussion notes from Area Maintenance Supervisor
- Discussion notes from Shift Supervisor.

The writer should study *all* of the above information items to gain familiarity with the machine and its problems.

WRITE DAILY INSPECTION ROUTE SHEETS (DIRS)

Certain preventive maintenance tasks should be performed on a daily basis. These tasks are normally of a very minor nature many of which can be performed while the equipment is running. These tasks represent a fixed, routine and repetitive workload. For this reason a different system is used to schedule this type of work than the system used for odd frequency preventive maintenance tasks.

Daily inspections are performed to satisfy ourselves that critical equipment is functioning satisfactorily and that situations aren't developing that might precipitate pyramiding failures and downtime. For example, if a certain pump was considered to be critical, it *might* be decided to inspect it every day as follows:

1. *Feel* all bearings for proper temperature
2. Check that gland leakage is not excessive
3. Check for abnormal vibration
4. Check oil level in bearings
5. Etc.

Each of the area maintenance supervisors should have the responsibility of developing *DIRS* on the critical equipment items in their respective areas. This task should proceed as follows:

1. Mark on the plot plan which of the items are in the specified area
2. Divide the area into the logical number of routes
3. Determine and mark on the plot plan the sequencing of inspections.

The following drawing illustrates this procedure (Figures and routes are hypothetical rather than actual).

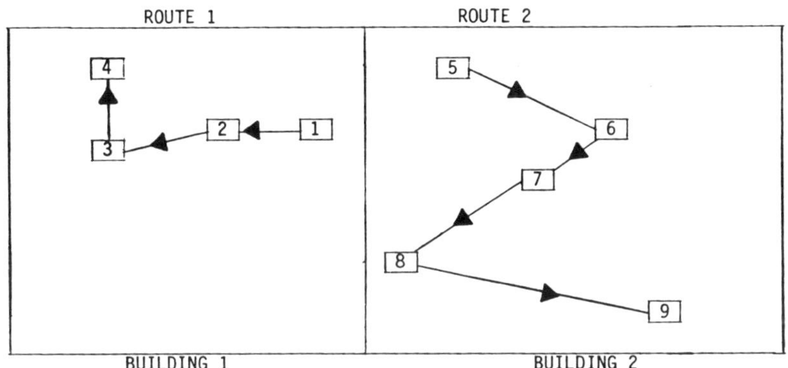

The numbered squares show that nine of the critical equipment items are located in Buildings 1 and 2.

The next step is to write inspection instructions on the form shown in Figure 2-15.

The first *Daily Inspection Route Sheets* developed should be construed as a "first-cut". In keeping with the intent of the "first-cut" effort, the route sheet writer should restrict his instructions to *obviously* essential items. Marginally effective instruction items can be given more careful thought when the writer assigned to a specific equipment item is working on the odd frequency instructions.

WHO SHOULD PERFORM DAILY INSTRUCTION SHEET TASKS?

The following personnel are candidates for performing daily inspection route sheet tasks:

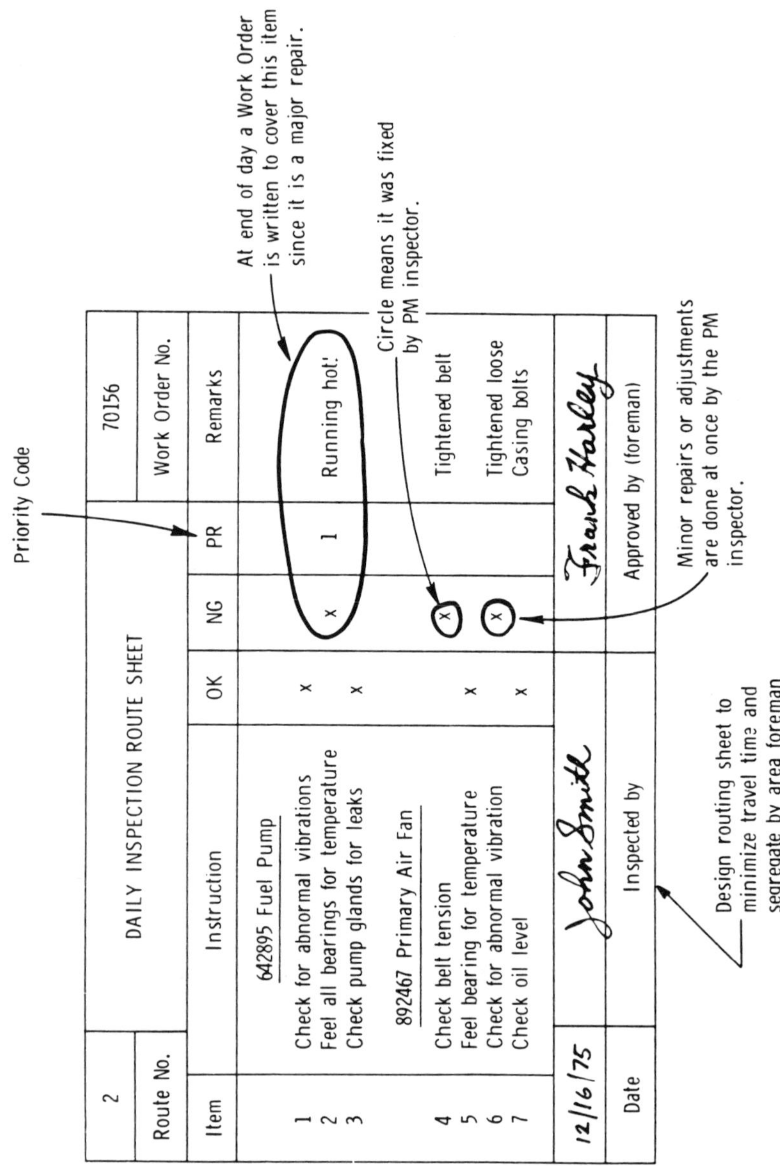

Route No.	DAILY INSPECTION ROUTE SHEET				Work Order No.
2					70156
Item	**Instruction**	**OK**	**NG**	**PR**	**Remarks**
	642895 Fuel Pump				
1	Check for abnormal vibrations	x			
2	Feel all bearings for temperature		x	1	Running hot!
3	Check pump glands for leaks	x			
	892467 Primary Air Fan				
4	Check belt tension		(x)		Tightened belt
5	Feel bearing for temperature	x			
6	Check for abnormal vibration		(x)		Tightened loose Casing bolts
7	Check oil level	x			
12/16/75	John Smith		Frank Harley		
Date	Inspected by		Approved by (foreman)		

Figure 2-15: Daily inspection route sheet.

1. Production supervisors
2. Maintenance foreman
3. Production workers
4. Maintenance workers.

Depending on the circumstances, one or the other of these groupings may be the more logical choice. For example, *if* a production worker is specifically assigned to watch a piece of equipment, and *if* the daily inspection sheet is within his technical capability, and *if* the task does not interfere with his normal assignment, then *he* may be the most logical choice.

It is up to the instruction writer to make recommendations concerning this matter. There may be cases where the most economical choice is to break up route sheet tasks to several of the four groupings.

PERFORMING DAILY INSPECTION ROUTE SHEET TASKS

The person making the inspection performs the work in the order shown on the instruction form.

After inspecting the item he puts an "x" either in the OK or NG (no good) column, as appropriate.

If the item is "NG" and the fix required is minor and within the skill limitations of the inspector, he shall fix the item on the spot. By "minor", we are referring to minor adjustments, such as, tightening nuts or bolts, tightening gland seals, adding oil, adjusting belts or other items, etc. It is not intended that he spend more than 2 hours per day on "fixing" things.

Whenever an item is marked "NG", an explanation is given in the remarks column to clarify the problem. If a "fix" is made, a circle is drawn around the "x" in the "NG" column signifying that problem was corrected on the spot. Discrepancies that are beyond the inspectors capability to fix are left uncircled. In this case the inspector indicates his appraisal of the seriousness of the problem by inserting a priority code number.

Examples of the various cases are shown in Figure 2-15.

At the end of the day, the inspector dates and signs the sheet indicating completion of the task. This is a very important step. Inspectors should be made to understand that signing off as complete, items that were not performed, is a cause for dismissal. Performance against these tasks should be periodically audited by the area foreman.

If a discrepancy is considered to be an emergency situation, the inspector should notify the area foreman or the maintenance office immediately.

After the foreman receives and notes discrepant items on the route sheet, the sheet is turned over to the chief planner scheduler who shall:

1. Write special work orders for the uncircled discrepant items
2. Plan and schedule these work orders
3. File route sheets chronologically by area and route number. These shall be held for a year for auditing and analyses purposes; then discarded. The preventive maintenance supervisor should review the check marks during the annual audit. Instructions with *zero* "NG" check marks should be considered for cancellation. Instructions with *very* frequent "NG" checks should be candidates for revision.

CHARGING ROUTE SHEET WORK

A blanket work order should be written for each daily inspection route. All work in this category should be charged to the appropriate blanket work order.

PREVENTIVE MAINTENANCE INSTRUCTIONS (PMI's)—GENERAL

All frequencies above daily should be specified and scheduled on preventive maintenance instructions. This work will constitute the major portion of the preventive maintenance system workload.

Since there will be many different equipment items, frequencies and instructions, the scheduling task can be complex and confusing unless a special scheduling system is designed. Since most companies now have computers, the system described in the following paragraphs is a computerized system which utilizes a simple sorting program and is designed purposely to simplify the scheduling function.

The writing of preventive maintenance instructions for the critical equipment items starts with an appraisal of the existing situation. If there are no breakdowns or an insignificant number, concurrently with low or reasonable maintenance costs, there may be no need to write preventive maintenance instructions.

However, if breakdowns are high, concurrently with excessive maintenance and production lost time costs, preventive maintenance instructions are probably justified.

The first question that a preventive maintenance instruction writer should ask is "What am I trying to achieve?"

The answer to this question can be found in Figures 1-2 and 2-14. We are trying to achieve on each piece of equipment, the lowest combined cost of maintenance and production loss cost caused by downtime. (Or the minimum downtime point or zero downtime).

The next question is, "suppose there is never or seldom a production loss when an equipment item breaks down, what then?" In this case, we are looking for the least combined cost of breakdown and preventive maintenance.

The writing of intelligent preventive maintenance instructions is the most important and difficult assignment in the preventive maintenance program.

The most common fault is to underestimate the skill required to write preventive maintenance instructions. Preventive maintenance instructions improperly written can scuttle a preventive maintenance program by increasing costs, and manpower requirements without off-setting decreases in production lost time costs. The knowledge required to properly write preventive maintenance instructions is acquired by a lifetime of exposure to maintenance problems and their solutions. The preventive maintenance instructions should give full consideration to the philosophies espoused in Chapter 13, concerning positive versus inspection, mandatory versus nonmandatory, pyramiding versus nonpyramiding and Figure 2-14, when writing preventive maintenance instructions.

Furthermore, the ratio of overhaul cost to production loss cost should be considered. Large ratios should discourage major overhauls. Small ratios should encourage overhauls.

When this ratio is large, the emphasis should be on examining historical failure records and writing preventive maintenance instructions slanted towards specific highly repetitive failures rather than major overhauls.

One must also realize that preventive maintenance outage costs, potential production loss costs from lack or addition of preventive maintenance work and the specific content of the preventive maintenance instruction are inextricably interwoven. Instructions which ignore this relationship are doomed to failure.

As the preventive maintenance instruction writer writes *each* instruction, he must ask the question, "Is this instruction going to reduce the combined cost of maintenance and production loss?" If the answer is *no*, then the instruction should not be written, unless safety is a factor.

In most cases, it will be fairly obvious whether or not a preventive maintenance instruction is worthwhile. In some cases, it may not be. As a matter of fact, the answer of whether or not it will pay off may be indeterminate. In this case, the preventive maintenance instruction writer must make an educated guess. The failure analysis chart, Figure 2-3, will give the final answer of whether or not the guess was correct or almost correct.

This word of caution concerning cost impact is given because when a preventive maintenance instruction is implemented, there is an immediate cost increase. We still have the existing costs until the preventive maintenance instruction has a chance to exert its cost reducing effect.

For this reason the preventive maintenance instruction program should be implemented gradually. When the preventive maintenance instruction on the first of the critical machines is written, it should be immediately implemented. The others should be implemented as they are written.

When the preventive maintenance instructions on all critical machines are implemented, there should be a pause for at least one month, during which time the preventive maintenance supervisor should assess the improvement process. When he is satisfied that there has been an actual cost reduction, he can start implementing machines in a lessor criticality category.

In addition, the preventive maintenance instruction writer should have a thorough understanding of the principals of failure analysis given in Chapter 10. These principals affect not only the writing of preventive maintenance instructions but also the revisions, updating and improvements to preventive maintenance instruction which are a natural part of the maintenance analysis process.

PREVENTIVE MAINTENANCE INSTRUCTION COMPUTER INPUT FORM

The form to be used in writing preventive maintenance instructions is shown in Figure 2-16.

Notice that the form is segregated for three trades:

- Mechanical
- Electrical
- Lubrication

Other types can be added, if necessary.

Sheet 1 of 1

682958	P. M. INSTRUCTION	21569
MACHINE NUMBER		WORK ORDER NO.
MACHINE DESCRIPTION		COST CENTER NO.

ITEM	INSTRUCTION	MECHANICAL					ELECTRICAL					LUBRICATION				
		W	M	Q	S	A	W	M	Q	S	A	W	M	Q	S	A
1	Inspect Oil Level, add Mobil XYZ, if necessary.	.2														
2	Renew Inlet Filters.			.2												
3	Open Fan Inspection Plate; inspect and report findings.					.3										
4	Renew Belts.					.3										
	SCHEDULING CODE NO.															
	TOTAL HOURS	.2		.2		.6										

Figure 2-16: PMI computer input form.

Each of the three trades have columns for the various frequencies which are:

W—Weekly

M—Monthly

Q—Quarterly

S—Semi-Annually

A—Annually

A rough estimate should be entered as shown opposite each instruction item. The purpose of this estimate is to give an approximate idea of workload for scheduling purposes. The estimate should be given in hours and tenths of an hour.

A blanket work order number should be assigned to each preventive maintenance instruction. *All* work performed against the preventive maintenance instruction should be charged to the blanket work order number. (The machine number put on the blanket order will be the specific machine for which the preventive maintenance instruction was written).

If it is decided to have an oiler perform a mechanical inspection, then the inspection should be listed under the lubrication column. In a similar manner, if an electrical inspection is to be performed by a mechanic, it should be listed under the mechanical column. This is done for scheduling purposes.

LOAD BALANCING

It is advisable to spread the preventive maintenance instruction workload as evenly as possible throughout the year in order to avoid disruption of other maintenance work. The scheduling code number is a computer tool for accomplishing this. A space is provided on the preventive maintenance instruction sheet for this code, see Figure 2-16.

There are other considerations affecting the date chosen to perform the preventive maintenance instruction, i.e.:

- Seasonal shutdown of parts or all of the plant
- Need to leave one redundant unit operating while the other is given its preventive maintenance instruction
- The logic of taking certain groupings of equipment out of service at the same time
- Other reasons of a similar nature.

It is advisable to let the computer do the load balancing operations, since practically all plants these days have, at least, business computers and since manual balancing is quite tedious. However, the decisions on the considerations mentioned above, can only be made by humans, and are mainly common sense decisions which must be tailored to each plant.

The load balancing function is facilitated by use of a table as shown in Figure 2-17. A separate table is made for each trade. This table can be filled in by hand or inputed to a computer which can then print the table. The steps in using the table and schedule code number are:

Step 1: Enter the hours and equipment number from the total column of the preventive maintenance instruction for the first machine. Notice that the table has the 52 weeks in the year in vertical columns and equipment numbers in horizontal columns. The total hours are continuously updated as each new machine is added to the table.

Referring to the preventive maintenance instruction Figure 2-16, we notice that there is a weekly task of .2 hour. Therefore, we enter .2 hour for each of the 52 weeks. Then there is a quarterly task of .2 hour. Now we have a choice of when we perform this task. Let us choose the 1st, 14th, 27th and 40th weeks. The .2 hour should be entered into the chart opposite the appropriate week as shown. Finally, let us choose the 52nd week to perform the annual task and enter the amount as shown.

Step 2: Select the scheduling code from Figure 2-18. We can see from this figure that there are the following choices of codes depending on the task frequencies:

Frequency	Number of Choices
Weekly	1
Monthly	4
Quarterly	13
Semi-annually	26
Annually	52

The code number is arranged as follows:

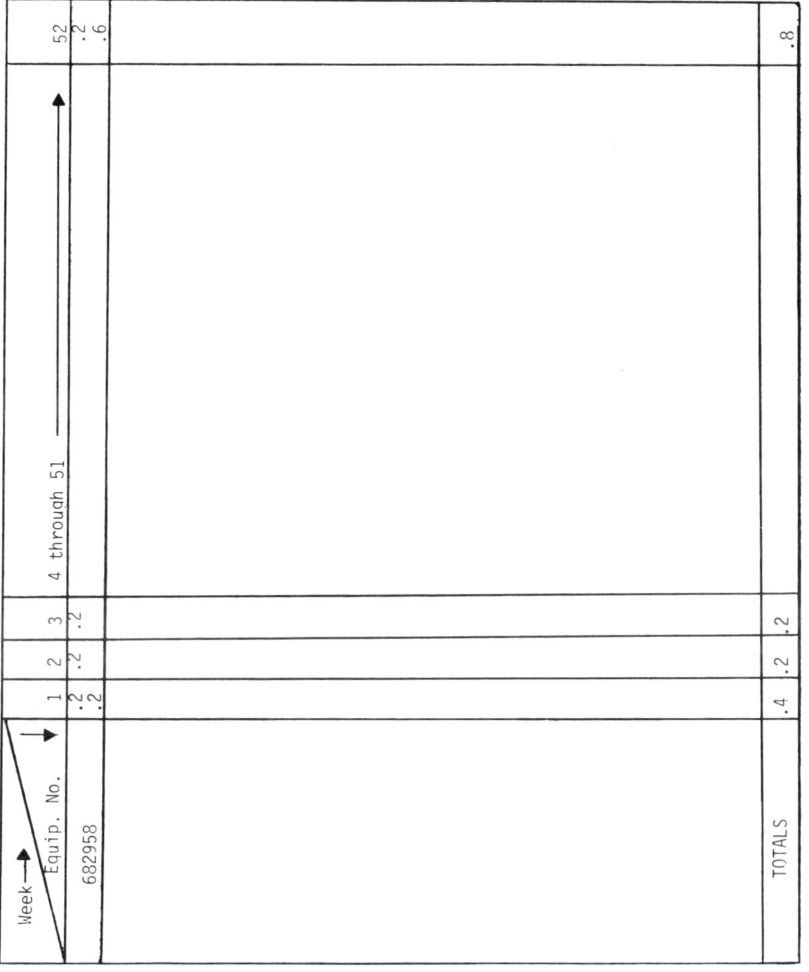

Week → / Equip. No. ↓	1	2	3	4 through 51 →	52
682958	.2 .2	.2	.2		.2 .6
TOTALS	.4	.2	.2		.8

Figure 2-17: PMI load balancing chart (mechanical).

WEEK	SCHEDULING CODE				
	WEEKLY	MONTHLY	QUARTERLY	SEMI-ANNUALLY	ANNUALLY
1	1	2	06	19	45
2	1	3	07	20	46
3	1	4	08	21	47
4	1	5	09	22	48
5	1	2	10	23	49
6	1	3	11	24	50
7	1	4	12	25	51
8	1	5	13	26	52
9	1	2	14	27	53
10	1	3	15	28	54
11	1	4	16	29	55
12	1	5	17	30	56
13	1	2	18	31	57
14	1	3	06	32	58
15	1	4	07	33	59
16	1	5	08	34	60
17	1	2	09	35	61
18	1	3	10	36	62
19	1	4	11	37	63
20	1	5	12	38	64
21	1	2	13	39	65
22	1	3	14	40	66
23	1	4	15	41	67
24	1	5	16	42	68
25	1	2	17	43	69
26	1	3	18	44	70
27	1	4	06	19	71
28	1	5	07	20	72
29	1	2	08	21	73
30	1	3	09	22	74
31	1	4	10	23	75
32	1	5	11	24	76
33	1	2	12	25	77
34	1	3	13	26	78
35	1	4	14	27	79
36	1	5	15	28	80
37	1	2	16	29	81
38	1	3	17	30	82
39	1	4	18	31	83
40	1	5	06	32	84
41	1	2	07	33	85
42	1	3	08	34	86
43	1	4	09	35	87
44	1	5	10	36	88
45	1	2	11	37	89
46	1	3	12	38	90
47	1	4	13	39	91
48	1	5	14	40	92
49	1	2	15	41	93
50	1	3	16	42	94
51	1	4	17	43	95
52	1	5	18	44	96

Figure 2-18: Scheduling code.

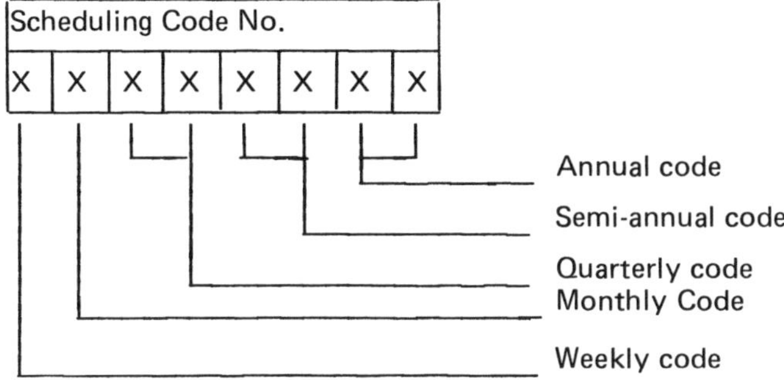

Since we have already determined which week each task is to be performed, we can then get the code number from Figure 2-18 which is:

1 - 0 - 06 - 00 - 52

The zeros in the monthly and semi-annual spaces indicate that there are no instructions for these frequencies for the first machine.

Step 3: Each week the computer will observe the scheduling code numbers "valid" for that week, since it has a copy of Figure 2-18 stored in its memory.

For example, for the 1st week of the year the "valid" code numbers are:

1 - 2 - 06 - 19 - 45

The computer then compares the scheduling code for the specific machine and determines where there is a match. Using the preventive maintenance instruction code previously determined and the 1st week, we have:

Valid numbers	1	-	2	-	06	-	19	-	45
PMI code	1	-	0	-	06	-	00	-	52
Match ? Yes or No	Yes		No		Yes		No		No

In this case there is a match for the weekly and quarterly instruction, therefore, the computer will print out instructions to do these tasks as shown in Figure 2-16.

Step 4: When the second item of equipment is ready to be coded, we consider the load spreading features of the scheduling code and its choices. Now when we select the various numbers to make-up the schedule code, we will select codes to spread the load as evenly as possible throughout the years—52 weeks. This is advisable since we would like to have a fixed size crew to handle preventive maintenance to the extent possible. More will be said about scheduling preventive maintenance in Chapter 16.

SCHEDULING

Each week on Friday afternoon, the computer shall print a set of instruction cards which specify the preventive maintenance tasks that must be done the following week. Figure 2-19 shows the form to be used by the computer. The cards should be printed in duplicate and segregated by:

- Area Code
- Mechanical Work
- Electrical Work
- Lubrication Work

In other words, no single card should contain instructions for more than one area or trade category.

The two sets of cards properly segregated should be distributed to each area foreman, so that he can plan his preventive maintenance work for the next week.

The foreman should retain one set of cards for a follow-up. The other set should be given to the workers so they can perform the work.

When the worker finishes the task specified on the card, he should initial the box labeled "complete" signifying that he has completed all tasks on the cards.

If he completes each task exactly as specified, he puts a check mark opposite the item. If unanticipated trouble was found or if he wishes to comment on an item, he puts an "x" mark opposite the item and puts his comments on the reverse side with an appropriate item number reference. For example, in the "x" marked item in Figure 2-19, the comment might read:

> 3–Fan impeller badly corroded. Should be replaced.

When the hourly worker returns the signed off card to the foreman, he matches it with his copy; if the card contains all check marks, he throws away both copies. (After noting time card charges).

If the mechanics card contains an "x", the foreman throws away his copy, notes the discrepancy and forwards the mechanic's card to the Chief Planner Scheduler for action.

Any cards left over with the foreman indicate that a card given to a mechanic has been lost or destroyed. The foreman must take appropriate action in the case.

More details on the scheduling of preventive maintenance instruction *after* the cards are received by the foreman are given in Chapters 15 and 16, Planning and Scheduling.

COMPUTER FUNCTION–REPORTING

In summary, the computer will record and print out costs for preventive maintenance work in the same manner as other work, by means of time cards and work orders. The following table summarizes the various cases of preventive maintenance charges:

Type of PM Order	Remarks
PMI work	Blanket work orders charged to specific equipment
Daily inspection route sheets	Blanket work orders charged against maintenance areas
Discrepancies surfaced by PMI's or DIRS's	Specific work orders charged against specific equipment

P M I	682958	Primary Air Fan	F.W. F.W.	M	26	2	1569
	Mach. No.	Description	Complete	Trade	Week	W.O. No.	

No.	F	Instructions	Hr	√ x
1	W	Inspect oil level; add mobil XYZ, if necessary	.2	√
2	Q	Renew inlet filters	.2	√
3	A	Open Fan inspection plate insp.; report findings	.3	x
4	A	Renew belts	.3	√
		see remarks for items marked "x" on reverse side		
		Total		

(BACK)

3. Fan impeller badly corroded. Should be replaced.

Figure 2-19: PMI scheduling card.

PREVENTIVE MAINTENANCE WORK PERFORMANCE

Concerning the performance of preventive maintenance work, the most important thing to remember is that the mechanic is in the most advantageous position to judge the quality of preventive maintenance instructions. He should not only perform each preventive maintenance task as directed, but should also be *specifically* instructed to maintain an open mind in judging the effectiveness of each preventive maintenance instruction. Some of the items for him to look for are:

- Do frequencies appear to be too high, too low, or just right?
- Are repetitive breakdowns occurring on parts or systems *not* covered by preventive maintenance?
- Should certain inspection tasks be changed to positive tasks or vice-versa?
- Should certain tasks be eliminated?

Foremen should periodically cover these points with the mechanic in a mutual discussion designed to draw out ideas for improvement from the mechanic.

AUDITING OF PREVENTIVE MAINTENANCE WORK

Foremen should periodically audit the mechanics performance of preventive maintenance tasks to ensure that all tasks are completed as signed off.

Signing off on an incomplete task is a serious *offense.* This point should be made quite clear.

Foremen should remind mechanics that they are expected to make minor adjustments and repairs while in the process of performing work.

The audit should be performed as frequently as is necessary for the foreman to have confidence that the mechanic is not only performing each preventive maintenance task scheduled, but that he is actively engaged in analyzing preventive maintenance tasks and discrepancies with a goal of constant improvement to written preventive maintenance instructions.

Part 3

PRODUCTIVITY CONTROL

14

Factors Affecting Productivity

GENERAL

Some of the factors affecting productivity are:

- Crew size and mix
- Material availability
- Tools and equipment availability
- Travel time
- Job planning
- Outage planning
- Working conditions
- Job method
- Employee morale
- Scheduling
- Employee skills

Each of these factors must be controlled basically as follows:

1. Establish policy
2. Measure performance against policy
3. Take corrective action when deviation between policy and performance occurs

The manner in which the control is used will vary for each of the eleven factors affecting productivity, but the principle will be the same. An example of this control cycle for the productivity factor of "crew size" might be as follows:

> *Policy:* Packing of pumps under 50 HP shall be performed by one man.
>
> *Measurement:* On work order X42AB, 3 men were used to pack a 10 HP pump.
>
> *Corrective Action:* Foreman was advised of his error and requested to comply with policy.

In the beginning there may be no pre-set policy on certain productivity factors, however in this case the factor should still be periodically audited on a spot sampling basis, and policy set after the fact. An example of this might be a measurement of "working conditions". An audit reveals complaints from the men and the union about the shop area which was very dirty and covered with oil, chips, dust and trash. This resulted in a policy being established to have machine operators sweep their waste material into the aisle and to have janitors collect the aisle dirt. Work sampling should be done at least annually in each maintenance area. Past work sampling studies should be used for comparison purposes after the first audit to determine if improvement has been made in the productivity factor. Suggestions for controlling each of the eleven productivity factors are contained in the following paragraphs.

CREW SIZE AND MIX

The crew size and mix of labor on a job is one of *the major* factors influencing productivity.

An example will illustrate this point. Changing the gland packing on a small pump is a one man job for the simple reason that there is only room for one man to work. If you put three men on the job, the productivity will be 33 ⅓% maximum. It's as simple as that.

Foremen should be constantly conscious of the crew size problem. They should make certain that excess personnel are used on backlog jobs rather than for "loading" existing jobs beyond their economical crew size.

Policy should particularly be established on highly repetitive jobs.

Auditing of this problem is simply a matter of spot checking jobs in the field. The correct crew size is normally quite obvious.

MATERIALS–GENERAL

The availability of material required to do each job is a *major* factor affecting productivity. More specifically, it's the *time* required to obtain material that principally affects labor productivity. This time is non-productive. There are four systems that are employed in cutting down on maintenance material acquisition time:

- Effective stocking of parts
- Bench stock
- Kitting
- Effective procurement

Each foreman has the resonsibility of adding new items to the stockroom as required by virtue of changing conditions. The stockroom system should take care of items that are no longer being used.

MATERIALS–BENCH STOCK

The free issue bench stock principle recognizes that low cost, fast moving stock items frequently account for a major percentage of the number of items used but only a minor percentage of the total inventory dollar value. For example, in one plant, 80% of the line items account for only 4% of the dollar value of stock items issued and/or stocked. For these type items the cost of processing issue slips may be very uneconomical. Recognition of this fact has led many to the bench stock principle, which provides for these materials to be placed in open bins available to the mechanic in a place convenient to their use point. This saves processing cost and mechanics time.

The following procedure can be used to determine the most "economical" number of bench stock items. The most "economical" refers to the lowest cost of the sum of inventory and material acquisition time costs.

a. Provide a special computer report (one time) that shows a year's history on *maintenance* stock items issued. The tabulation should be in the format shown in Figure 1-4 with stock items and cost substituted for equipment numbers and cost.

   This tabulation will permit curves to be drawn similar to those shown in Figure 3-1.

   The shape of the curve will indicate the extent to which the free issue stock principle may be employed. A curve similar to the one shown in curve A would encourage a large number of free issue items whereas curve B would discourage free issue items.

b. Convene a meeting of all maintenance foremen. In this session prepare, with everyone's help, a list of highly repetitive maintenance jobs. Divide the number of jobs on this list evenly by foremen and have them prepare a list of highly repetitive stock items used on these jobs.

c. Those items which represent fast moving low dollar value parts observed on the tab run (representing a large number of items but a very small number of dollars) and which also show up on the list of highly repetitive items used for highly repetitive jobs should be made bench stock items.

d. When contemplating where to locate the bench stock, the following alternates should be considered depending upon the circumstances:

   - At the equipment: When the item is used for specific equipment.

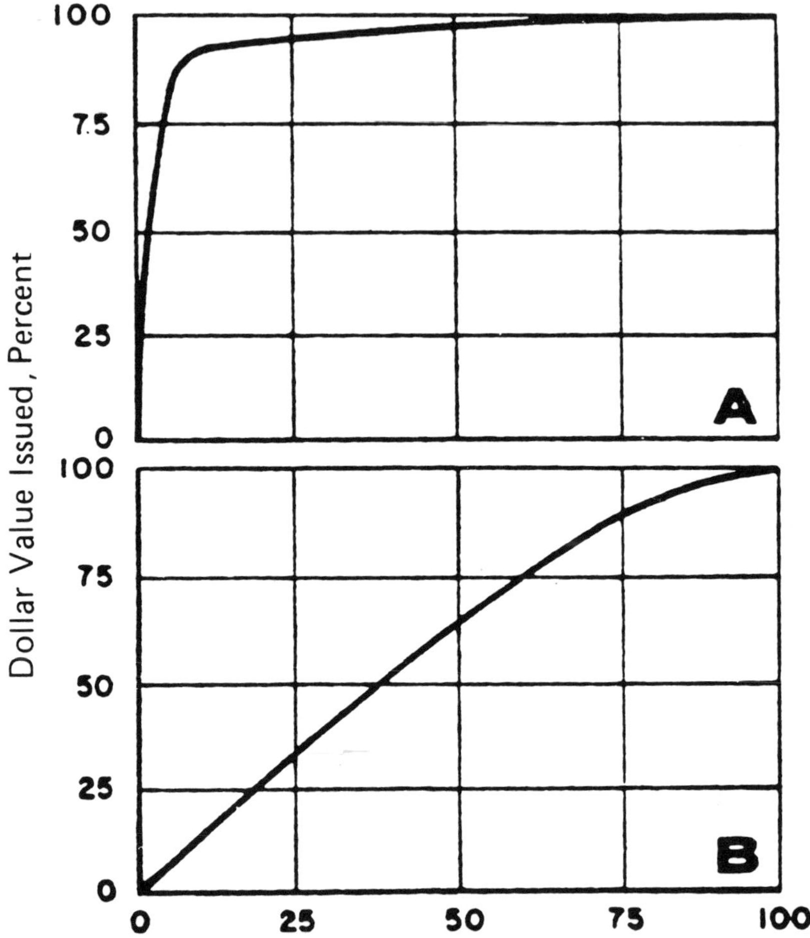

Figure 3-1: Bench stock indicators

- At the center-of-gravity-of-use point: When the item is used throughout the plant. This could include one or more locations.
- In the storeroom: In a non-secured area.

e. In addition, a decision will have to be made concerning who will replenish the bins—stores or maintenance personnel. The location and replenishing decision must be decided upon the basis of an economic analysis.

MATERIALS–KITTING

Kitting is defined as the process of obtaining all materials for a specific job in advance of the job start date. A fully kitted work order enables the job to proceed without interruption due to material unavailability. Kitting therefore increases productivity.

Construction, modification, rearrangement, and major maintenance work can be conveniently kitted due to the fact that this type of work can normally be planned well in advance of its need date. Because of this and the beneficial effect kitting has on labor productivity, this type of work should be kitted.

Referring to the list of maintenance priority codes in Chapter 9, normally Priority Codes 3 and 4 should be kitted. Priority Codes 1 and 2 (Emergency and Urgent Work) will normally not be kitted due to the lack of time. Priority Code 5 work will normally not be kitted because of the marginal justification implied in this code. This work may, however, be kitted upon approval of the maintenance manager.

Kitting should be accomplished by utilizing the following procedure:

Maintenance Responsibility

(a) After the work order has been prepared and approved, the planner for the particular job should prepare a bill of materials similar to that shown in Figure 3-2.

(b) Prepare purchase requisition for all outside purchases. (A note shall be placed on the

BILL OF MATERIALS

Date 11/10/87 Due Date 11/30/87 Work Order No. 672

Item No.	Quantity	Description	Stock No.	Date Received
1	20	2"-90° SCH 40 GALV. ELBOWS	1A 256	11/22/87
2	40	2" CLOSE NIPPLES GALV SCH 40	1A 205	11/22/87
3	30	2" GALV TEES SCH 40	1A 212	11/22/87
4	1	PUMP-MODEL 240-160GPM AT 100 PSI	—	—
5	10	2" BRASS GATE VALVES #150	—	—
6	4	2" SOLENOID VALVES MODEL 15	—	—
7	1	2" CONTROL VALVE MODEL 10	—	—

Sheet 1 of 2

Figure 3-2: Bill of materials.

purchase requisition to have the purchased item tagged for the work order.)

(c) Send bill of materials to warehouse.

Warehouse Responsibility

(a) Establish manila file folder for each work order which is to be kitted.

(b) Insert copies of bill of materials, purchase requisitions and purchase orders in folder.

(c) Write stock requisitions and fill all stock items on bill of materials.

(d) Designate kitting area for work order and mark location on work order kitting folder.

(e) Deliver stock items to designated kitting area.

(f) Receive, check off, and deliver outside purchases to kitting area.

(g) When entire bill of materials is checked off as complete, call Maintenance and notify them to pick up completed kit.

MATERIALS–ACQUISITION TIME

Because material acquisition time is a major item of nonproductive maintenance labor, it is worthwhile to conduct a periodic audit of this time to have some feel for the success or failure of improvements and to determine when the time spent is no longer of concern. The following procedure is recommended.

- That the maintenance planner conduct a semi-annual audit of this problem area.
- Each audit shall last one week.

During each 7 day audit period, all members of the maintenance staff shall record, on a daily basis, all of the time they spend obtaining materials. "All members" inlcude:

- *All* supervision
- *All* staff
- *All* craftsmen

It excludes:

- Engineers
- Operating personnel
- Warehouse personnel
- Purchasing personnel

All time spent in obtaining materials includes:

- Time waiting for materials
- Time traveling for materials
- Time transporting materials
- Time identifying materials
- Time substituting materials
- Time locating materials
- Purchase Requisition preparation time
- Approval time
- Non-productive time caused by being held up by other labor without proper materials; wrong materials delivered; wrong materials ordered; material stock depleted

Time should be recorded in minutes.

The maintenance planner should prepare an audit report showing:

- Total hours and minutes worked that week
- Total hours and minutes spent in obtaining materials
- Percent time spent in obtaining materials

This report will indicate the seriousness of the problem. Future reports will measure the success or failure of corrective action.

TOOLS AND EQUIPMENT AVAILABILITY

We must be concerned about having proper tools and equipment on all jobs but most particularly in the following cases:

- Highly repetitive failures.
- Failures on critical production equipment.

In these cases, the highly repetitive failures surfaced by the Failure Analysis Report, Figure 2-3, should be analyzed carefully to see if any special tools, equipment, jigs or handling devices can be designed or purchased in order to reduce the repair time.

TRAVEL TIME

Travel time should be audited and controlled in a manner similar to that described for material time, if it is considered to be a problem.

JOB PLANNING

In order for a maintenance crew to be able to proceed with a job, they must have "instructions" or a "work plan." If the work isn't planned before the crew is sent out to do a job, it must be planned during the job. There are several things wrong with the latter course of action:

1. Preplanned jobs are normally planned by one person. Jobs planned during the job usually tie up the entire crew plus the foreman in the planning function. It would not be unlikely to have a planner take ½ hour to plan a job as compared to an 8 man crew take all day or 64 hours to plan a job they were sent out to do without preplanning.
2. There is much more likelihood to run into a stumbling block like inability to get certain parts, or an outage, etc., with the result that

considerable idle time may be suffered. As a matter of fact, the idle time may amount to 100%, since the entire crew frequently is sent to another job because of the lack of material.

3. Planning done on a "surprise" basis usually doesn't match the thoroughness and accuracy of planning done by an experienced planner before the job is started.

CASE STUDY 3A: The Missing Bracket

While working in an aluminum extrusion plant, a system for automating a critical machine was designed. I was excited about the design and could hardly wait to see it being installed and working, because it promised to achieve large manpower savings.

On the day installation was supposed to start, I was out at the machine at 8 a.m. I was disappointed to see the crew of 6 mechanics and one electrician arrive one hour late at 9 a.m. and made a mental note to talk to their foreman about this waste of 7 manhours.

The crew gathered around the prepared drawings. They agreed that the starting point of the job had to be the installation of an electric solenoid operated multi-ported hydraulic control valve that was the heart of the system. After this was installed then the mechanics could start running the many hydraulic lines and the electrician the wiring. "Where is the bracket to support the control valve?" asked the lead mechanic. Another said, "There isn't one shown on the drawings."

Startled, I myself looked at the drawings and apologized for forgetting to design a bracket so that it would have been kitted and ready before the crew came out to the job. The lead mechanic told one of his men to get a welding machine, cutting torch and a length of 2″ × 2″ × ¼″ angle iron so that a bracket could be made. The men sat around and gossiped while waiting for the mechanic to come back with the welding machine and angle iron.

Since the mechanic wasn't back by 11:30, they all left for lunch. At 12:30 they were all back with the welding machine and angle iron. It only took 1 hour to make the bracket and the crew was now back at the starting point.

It would have taken me only 10 minutes to sketch a bracket on the drawings.

As a result of this omission the crew was idle for 3 ½ hours for a total lost time of $3.5 \times 7 = 24.5$ hours. I learned a lesson from this minor incident that I shall never forget! DON'T SEND CREWS OUT ON A JOB WITHOUT COMPLETE PLANNING IN ADVANCE!

OUTAGE PLANNING

The manner in which equipment outages are handled affects labor productivity.

Intelligent handling of this problem is complex. Suggestions for the proper handling of outages is given in Chapter 10.

WORKING CONDITIONS

Working conditions affect not only productivity but also:

- Morale
- Production loss
- Health
- Safety

Workmen should be instructed to clean up after each job as part of their contribution to keeping the plant clean.

Foremen should see to it that their shops are not only efficient, but also have the visual appearance of being efficient.

This is important from a morale standpoint, and to promote a feeling of confidence on the part of customers of the shops.

JOB METHOD

The method used in doing a job influences productivity. The priority given to a job method improvement study on a specific job should be proportional to:

- The productivity percentage of the job.
- The potential for improvement.

For illustration purposes let us assume that a work sampling study has been made on a group of maintenance workers doing a specific job and the results are as follows:

	Percent
Travel Time	15
Material, tool and equipment acquisition time	50
Idle time	10
Obtaining instructions	5
Working	20
Total	100

If in this case a quick investigation reveals a reasonable job method, quite obviously the improvement study should concentrate on the excessive Material Tool & Equipment Acquisition Time rather than on the job method.

Conversely, if a time study on a repair job showed:

	Percent
Travel	14
Material	14
Idle	10
Instructions	2
Working	60
Total	100

then a study of the job method might be warranted.

EMPLOYEE MORALE

Each supervisor has the responsibility of being concerned about the morale of his employees and for doing what he can to keep this factor high.

The art of doing this is the art of "motivation". It is the responsibility of the maintenance manager to direct a motivational program under the guidance of top management.

The end result of the program is to develop in the employee a feeling that the company:

- Recognizes him as an individual
- Recognizes him for his accomplishments
- Recognizes him as an important factor in the success of the company.

Great sophistication in motivational program development is not needed. A simple and very effective motivational device, for example, is to slap an employee on the back periodically and thank him for a job well done. The three types of recognition listed above must be communicated and communicated repeatedly. These motivating communications are like a food that a person must have to survive and be productive.

A company or supervisor who does nothing positive in the art of motivation must inevitably suffer low morale and low productivity.

CASE STUDY 3B: "A Day's Pay for a Day's Work"

At one plant, it soon became obvious that the morale of practically all maintenance workers was very low. Their productivity as determined by work sampling and spot audits was correspondingly low. After being there for a month I found many reasons for the low morale.

What was most shocking to me was that I could find no one responsible for, or aware of, a motivation program. I was particularly surprised to find that the director of personnel said he was not responsible for a motivational program nor did he know of anyone who was.

I reported my findings to president of the company who said, "I pay them a day's pay for a day's work and thats all the damn motivation they need, or are going to get from me." In other words, the president himself was responsible for the problem. It was recommended to him that he read "Theory Z" and other related books on motivation after failing to convince him of the importance of motivational programs. He, of course was irritated by this suggestion.

Later it was found that this attitude was shared by many top executives, which I'm sure is partly responsible for the traditional conflict between unions and management in the U.S.A.

SCHEDULING

The kind of scheduling system varies with the type of maintenance work. This is discussed in Chapter 16. If the scheduling system isn't intelligently designed, or if there is no system, productivity will be adversely affected.

This is true because proper scheduling is necessary to insure that manpower, instructions, material, tools and outages be brought together in optimum time relationships.

EMPLOYEE SKILLS

Quite obviously the skill of an employee affects his productivity. An employee who is skilled at a particular job can do it faster than one who is not and do it so that it won't immediately break down again. If there is a productivity problem with an employee, it is important to determine if it is a morale or motivational problem as opposed to a skill problem.

There are three action items required of each maintenance supervisor regarding this productivity factor:

1. Measure the skill level of all employees working for him.
2. Communicate this measurement to the maintenance manager.
3. Apply training effort commensurate with the individual's weakness.

A great part of skill weaknesses will be automatically taken care of by an apprentice program. However, specific training for particular individuals may at times be appropriate.

15

Planning

GENERAL

This section will discuss planning. Definitions of related functions follow.

The basic purposes of planning and scheduling maintenance work are:

- To improve the productivity of maintenance labor
- To perform work on schedule
- To plan for future requirements before they are needed on a crisis basis

Planning is the determination of all elements required to perform a task in advance of the job start time.

Scheduling is the specific time phasing of the planned elements together with orders to proceed with the work, and follow up and status reporting on job progress.

Work accomplishment includes briefing the crew on the job elements, giving orders to proceed with the work, performing the work, job coordination and checking the quality of work in progress, and at completion.

Planning can be divided into three basic levels:

- Daily job planning
- Short range planning
- Long range planning

Daily job planning concerns itself with the determination of all elements required to perform an individual task in advance of the start date.

Short range planning concerns itself with an annual plan of operation. This plan will specify how the maintenance force will operate during the coming year and will provide detailed plans for major overhauls and construction jobs. An important part of the short range plan will be vacation planning and plant shutdowns, if any.

Long range planning covers a period of five years or more and provides plans for future activities and for long range improvements to the maintenance function.

LONG RANGE PLANNING

Long range planning of maintenance requirements is closely allied with, and dependent on, long-range sales and production forecasts. Such planning is accomplished by staff planners who develop a long range program for the entire company. The planners, working with line executives of the operating divisions, outline, in effect, what is needed in the way of decisions today in order to reach certain goals five years from now. The management level of the staff long range planners is usually very high, reporting directly to the president or to a vice president. The long range plans are projected throughout management, and those affecting production influence the planning of the maintenance manager and others throughout the organization.

The primary purpose of long range planning is to keep the maintenance objectives, policies, and procedures updated in order that they will be in tune with the objectives of the company. This requires knowledge of sales and production forecasts and consideration of all the factors involved in long range production planning. In addition, maintenance planning requires projection of two particular factors which are very important to the maintenance organization:

(1) Changes in maintenance equipment and facilities requirements

(2) Changes in production equipment due to obsolescence, increased mechanization, automation, and other technological improvements.

Thus, the planning of maintenance requirements for the future involves changes within the maintenance department as well as planning for the maintenance work to be done in support of production in the future.

Finally, long range planning includes general plans and objectives for improving the operation of the maintenance function. This includes items such as training programs for craft workers, management development programs for future maintenance managers and foremen, methods improvement objectives, new maintenance requirements arising from proposed capital equipment acquisitions, equipment replacement plans, and long cycle equipment overhauls.

In summary, long range planning takes a look 5 to 10 years into the future and makes plans for the maintenance department to cope with coming events. Predicting that will happen in the future is done in several ways:

(a) Studying the long range plan of production to determine what impact it will have on maintenance;

(b) Extrapolative planning, that is, the extrapolating of present trends on into the future;

(c) Entrepreneurial planning, that is, the prediction of maintenance impacts caused by changes in the external environment. This covers anything and everything outside of the company any place in the world that may impact the company's operations. Admittedly there will not be too much of this type of planning taking place in the maintenance department but is important to top management.

Long range plans should be made out once per year by the maintenance manager. Progress towards long range plans should be reviewed on a quarterly basis. The maintenance manager should call upon members of his staff to submit ideas for the long range plan. These will not only provide valuable ideas to the maintenance manager, but will also make the staff feel that their input is wanted and therefore have a motivating effect.

Examples of long range plans are shown in Figures 3-3 and 3-4. Foremen developing long range plans will find the format in the examples helpful.

Foremen should avoid making long range plans that are too generalized and lacking in specifics, for example:

- "Improve the efficiency of the department", or
- "Improve communications", or
- "Reduce travel time".

This fault can be avoided by developing a plan that is as specific as possible and by listing all of the steps requried to implement the plan complete with estimated completion dates. The maintenance manager should follow up on the due dates for milestones in the plans.

SHORT RANGE PLANNING

The short range plan is the annual plan of operation. It consists of the following:

- A general plan of how all labor should be distributed throughout the various departments under "normal" circumstances.
- The detailed planning and scheduling of major events taking place during the year such as

Long Range Plan No. *1*

For

General Maintenance Foreman – *Smelter*

Date Initiated: *1-1-72*

Description: *Select, train and develop a replacement for position of General Maintenance Foreman – Smelter (in case incumbent leaves, dies or retires).*

Implementation Steps

No.	Step	Status/Remarks	Estimated Comp. Date	Actual Comp. Date
1	*With the help of the Personnel dept., select 3 candidates.*		*1-15-72*	
2	*Screen, interview and select one of 3 candidates.*		*2-15-72*	
3	*Study strengths and weaknesses of successful candidate. Develop plan for overcoming weaknesses.*		*3-15-72*	
4	*Rotate candidate through all smelter maintenance supervisory jobs including Planner/Scheduler's position.*		*1-1-74*	
5	*When sufficiently trained, use as acting General Foreman during vacations, sick leave, etc.*		—	

Figure 3-3: Long range plan–example No. 1.

Long Range Plan No. *2*

For

General Maintenance Foreman – *Mill*

Date Initiated: *1-1-72*

Description: *Reduce travel time for all maintenance units*

Implementation Steps

No.	Step	Status/Remarks	Estimated Comp. Date	Actual Comp. Date
1	*Develop procedure for measuring travel time*		*1-15-72*	*1-17-72*
2	*Brief foremen on procedure*		*2-15-72*	*2-1-72*
3	*Measure travel time (one week long study)*		*3-15-72*	*3-15-72*
4	*Meet with foremen and discuss study results*		*4-1-72*	*4-1-72*
5	*Write analysis and recommendations. Add recommendations to this plan.*	*Study concludes that problem will be solved by purchase of 4 golf carts*	*5-1-72*	*5-1-72*
6	*Purchase 4 Cushman gasoline engine driven golf carts*		*9-1-72*	
7	*Measure travel time (one week long study)*		*10-1-72*	

Figure 3-4: Long range plan–example No. 2.

major overhauls, large construction jobs, major preventive maintenance tasks and vacations.

The first item shows the steady state situation. The second item shows the major perturbations to the normal schedule. These should be plotted on an annual Gantt chart of activities to give visibility.

The basic purpose of the short range plan is to provide adequate lead time to plan all elements of very large jobs and to describe the normal distribution of maintenance labor by work type. Lack of such a plan will inevitably lead to excessive downtime and costs due to unavailability of labor, materials, tools, equipment, access and instructions.

DAILY JOB PLANNING–GENERAL

Studies of typical maintenance departments often show that less than 25–30 percent of the day is spent on productive work, with more time spent on non-productive activities. Maintenance management has tried to find the answer to this basic problem of manpower utilization. The answer involves many factors and management tools.

One of the most effective tools used to prevent such waste of time and dollars is *daily job planning.*

The production of even the simplest of manufactured parts requires a series of steps involving many people to carry the product from the original idea to design, selection of methods, tooling and materials, and coordination of purchasing, processing, finishing, and distribution. But the entire procedure for making a maintenance "product" is often carried out by one person, the maintenance craftsman, when, for example, he is sent out on a job to "see what the foreman wants in department 39 and take care of it". He designs the product, decides on methods and materials, secures tools and supplies, delivers and installs, and even decides what account to charge the job to! If we really want him to spend more time working at his trade, it seems only logical to provide a staff service to carry out some of these indirect operations. Stated simply, the way to avoid excess trips back and forth to the shop for manpower, tools, materials and equipment is to provide an accurate description of what is needed before the job is started. The way to decide how to do the work is to analyze the job *before* it starts. This is the primary function of maintenance job planning.

Daily job planning provides the following:

Job Instructions

How to do the job including drawings and specifications if required. It is important to understand that any design, plan or instruction omitted may result in lost time for maintenance labor several times greater than it would take to do complete planning. This is particularly true if the job requires a multi-man crew.

Material Required

The planner should provide a list of all materials required and write purchase requisitions for non-stock items.

Labor Required

An estimate of the labor required, broken down by trades and crew size.

Special Tools and Equipment

A list of special tools and equipment needed for the job, other than ordinary tools and equipment.

Safety Instructions

A listing of unusual safety hazards involved in accomplishing the work and what to do about them.

Required Completion Date

The planner should arrive at the required completion date by estimating the following time requirements:

(a) Design and/or planning time
(b) Work order approval time
(c) Material delivery time
(d) Shop backlog time
(e) Work accomplishment time

These should be added to the time the project was received to arrive at the required completion date.

This should then be compared with the requested completion date. If the calculated date is later than the requested date, the planner should negotiate with the requestor for a revision to the due date.

Work Orders

The planner should prepare necessary work orders.

DAILY JOB PLANNING—WHAT IS PLANNED?

All work should be planned in advance; even emergency breakdowns and trouble call work can be planned in advance as described later in this chapter.

DAILY JOB PLANNING—WHAT IS ITS PURPOSE?

The basic purpose of planning is to reduce costs. An unplanned job will almost always cost more due to inherent inefficiencies.

DAILY JOB PLANNING—WHO IS RESPONSIBLE?

First of all there should be only *one person* held responsible for planning *one job* regardless of the number of departments, trades, skills, men or engineering disciplines required to plan the job. Some *one* person has to:

(a) Determine the components of the job. (In designing a battery charging room, he must know that an explosion proof exhaust fan is required for exhausting hydrogen gas. If he doesn't know this, the fan will be omitted.)

(b) Coordinate between all people involved in planning the job.

(c) Assume *total* responsibility for the design and/or planning of the job.

Without the assignment of one person as "project engineer" on a job, all of the above three points will become problems, with the result that portions of the job may be overlooked, there will be problems in coordination and if several people are involved without one person in charge, it will be impossible to hold one person accountable. Each person will say "Oh I thought John was responsible for that." Of course these problems will not exist for a large number of jobs that would ordinarily involve only one planner.

Job planning may be performed by any of the following people depending on the particular organization:

- Maintenance foreman
- Maintenance staff
- Engineering staff
- Planner scheduler

Furthermore, the maintenance manager may, at his discretion, decide to plan certain jobs himself.

DAILY JOB PLANNING—SCREENING

Screening is the process of questioning each maintenance job request in order to establish a firm economic basis as well as validity of need. The objective here is to put a dollar sign on the routine decisions regarding "Make or buy?", "Is it necessary?", and "How do we do it?". Screening should include consideration of prefabrication instead of on-the-job assembly, purchased parts instead of parts made on the job or in the shop, replacement instead of repair, and contract service. In other words, this procedure helps to avoid overlooking a better or cheaper way to get the job done and in deciding if the work should be done at all.

Examples of excessive cost resulting from lack of screening can be found in almost any plant maintenance operation and include:

- Repair of fractional horsepower motors at more than the cost of new motors
- Fabrication of workbenches and cabinets in a maintenance shop at several times the cost of commercial products
- One-at-a-time machining of standard gears that are available off the shelf.

DAILY JOB PLANNING—PREPARATION OF WORK ORDER PACKAGES

The planner, when making out a work order, should determine whether or not there is enough space on the work order to fully describe the work. If there is not, then the description should be short and refer to a supplementary document. For example:

CASE 1

"Furnish labor and material to fabricate and install a new control system for Mill No. 1 in accordance with drawing No. 12345, sheets 1 through 4 of 4."; or

CASE 2

"Furnish labor and material to rebuild Mill No. 1 in accordance with attached specification sheets 1 through 9 of 9." The specification sheet should have the same title as the work order and be identified by the work order number.

In other words the description for work should be COMPLETE, either by specifying everything on the work order or by referring to a supplementary document.

In Case 1, if there was a separate sheet for a bill of materials, then the description shown would be INCOMPLETE. It could be made complete by referencing a bill of materials on the work order or on the drawing (since the drawing is already referenced on the work order). The same applies to any other auxiliary document. The problem to be avoided with this procedure is the loss of any auxiliary document necessary to the work order without any indication on the work order that it exists.

DAILY JOB PLANNING—PROJECT PLANNER CONCEPT

If a job is poorly planned, it is usually for one of the following reasons:

(a) One person did not assume responsibility for the entire planning job, with the result that portions of planning were overlooked.

(b) Project planner was not qualified to plan the job.

(c) Project planner was careless in planning the job.

(d) Project planner didn't have the time to properly plan the job.

The project planner concept is simple. It merely states that for one job there shall be only *one* planner responsible. He must assume responsibility for coordinating the activities of any other people helping him in doing the planning.

The planner *must* determine if he is qualified to do the planning on a specific job. If he is not, he should disqualify himself and refer the job to someone who is qualified. Normally the maintenance manager should make this determination. Part of being qualified is having the knowledge of all the components that are necessary for a complete job. For example, in designing a boiler installation, he must know that a high pressure cutout is required. If he doesn't know this, it will be omitted.

Finally, if the planner doesn't have the time to plan a job, he must be aware of the consequences:

- If he holds the job in abeyance, what are the consequences of the delay?
- If he plans to put the job into work without planning, what is the extent of lost productivity?

If either one of these consequences is unsatisfactory, he should:

- Hold up another job and plan the job under scrutiny, or
- Arrange for someone else to plan the job.

The extent that a job is planned will depend on the type of work involved. The following paragraphs comment on the extent of the planning function by work type.

DAILY JOB PLANNING–EMERGENCIES

Previously, emergency work was defined as unplanned work and this is basically true. Nevertheless, there is a certain amount of pre-planning that can and *should be* performed. Examples of this are:

(a) Listing, by each foreman, of typical emergencies that may take place in his area, giving emphasis to emergencies on critical equipment and emergencies involving a safety hazard.

(b) Writing of emergency procedures covering list (a) that cannot be handled by normal knowledge and skills of incumbent personnel within reasonable time frames.

(c) Training of personnel in procedures developed in (b); including, in especially critical situations, "Fire Drill" type of training.

(d) Preplanning of availability of materials either in stock in the warehouse or at the equipment for the purpose of reducing response time.

(e) Preplanning of availability of special tools, fixtures, and equipment for reducing response time. This should include consideration of design of special repair aids, especially on critical equipment. Consideration should be given to locating the tools, fixtures and other aids at the equipment.

DAILY JOB PLANNING—PREVENTIVE MAINTENANCE

Preventive maintenance work is by definition all preplanned.

DAILY JOB PLANNING—CONSTRUCTION, MODIFICATION AND REARRANGEMENT

All of this type of work should be completely preplanned.

DAILY JOB PLANNING—TROUBLE CALLS

Trouble call work or minor breakdowns is a type of work that is difficult to preplan. The obvious reason for this is that you don't know what to do until you arrive at the machine and diagnose the trouble. Further, it is frequently necessary to dismantle the machine and look inside it before you can diagnose the trouble.

Nevertheless, a certain amount of preplanning is possible. The following suggestions are made:

(a) *Troubleshooting Skills*

Troubleshooting requires the highest order of maintenance know-how. Each maintenance unit should train and develop a crew of experts in this field.

(b) *Troubleshooting Aids*

Certain items of equipment are easy to troubleshoot; others are complex and difficult. Trouble analysis charts should be made up and hung in plastic envelopes on the complex machines. These charts should follow the format:

- Trouble
- Probable cause
- Remedy

With these charts in his possession, the troubleshooting mechanic can quickly identify the trouble, its probable cause, and how to fix it.

Trouble analysis charts of this type are usually provided by the equipment manufacturer, but are of little value if hidden in some unknown location in a foreman's desk.

(c) *Troubleshooting Materials*

Some preplanning of materials can be done. For example, it sometimes pays to have a troubleshooting cart, or hand tote box, supplied with items typically used on trouble calls. Another suggestion is to store minor materials at the machine. For example: fuses, special nuts and bolts, packing, insurance spares or electrical brushes.

(d) *Troubleshooting Tools and Equipment*

It goes without saying that mechanics should have in their possession hand tools and instruments required for the job. However, thought should be given to other aids such as conveniently located ladders and special tools, etc. In maintenance centers located remotely from the central shop, consideration should be given to creating subcenters for the purpose of storing needed tools and equipment that will reduce total travel time.

DAILY JOB PLANNING—MAJOR MAINTENANCE

Major maintenance jobs or major overhauls should all be preplanned. As a matter of fact, most major overhauls should be part of the short range plan outlined earlier in this chapter.

Major overhauls can be divided into two categories:

- Known Work
- Unknown Work

This can best be illustrated by examining the overhaul plan for a reciprocating pump:

Known Work

(a) Completely disassemble pump

(b) Renew all broken nuts and bolts

(c) Renew all gaskets, packing and piston rings

(d) Clean all surfaces; remove rust and scale

(e) Check all valve springs and renew broken or defective springs

(f) Check all valves for proper seating

(g) Paint all surfaces as required with paint spec XYZ

(h) "Mike" cylinder for barrelshapeness and out of round

(i) Check shaft, sleeves, and bushings for excessive wear or scoring

Unknown Work

(a) As a result of item (f) above, valves may have to be renewed and/or ground in.

(b) As a result of item (h) above, cylinder may require reboring or renewal of liner.

(c) As a result of item (i) above, shaft and sleeves may require renewal, building up and/or honing.

The point to be observed in this example is that a major portion of the overhaul work can and *should* be preplanned and can be estimated.

A critical item to be preplanned in major overhauls is the coordination required with production to obtain a mutually agreed upon outage time which will allow all of the elements of planning to be accomplished and a reasonable amount of time to do the work. To do less is to invite increased costs and equipment downtime.

On very complex items of equipment, particularly where downtime for overhaul causes a production loss, it may pay to draw a "PERT" or "CPM" diagram for the overhaul to maximize coordination, and reduce confusion and lost time.

DAILY JOB PLANNING—SAFETY ITEMS

The safety department should conduct routine *scheduled* inspections of facilities and operations to insure compliance with company safety practices.

Violations to these practices shall be recorded by the safety engineer making the inspections.

The safety engineer should then prioritize each item in accordance with the priority system described in Chapter 9.

Finally, the safety engineer should make a list of prioritized safety violations. A separate list should be made for each foreman, showing only the items under his jurisdiction. The list should be sent to each foreman, and a copy to the maintenance manager with instructions to eliminate the safety hazards by Priority as follows: (An explanation for the reason for each item should be included in a "remarks" column. If the item represents a violation of code regulations, the code should be referenced.)

> *EMERGENCY—Priority 1*
>
> Safety items in this category constitute an existing and immediate threat to life and should

be acted upon immediately. The safety engineer should have the *authority* and *responsibility* to shut down the operation if in his opinion it is necessary to protect life. Emergency items should be reported immediately to the maintenance manager by phone. Modification or new construction work required to eliminate the hazard should be covered by "after the fact" work orders.

URGENT—Priority 2

Safety items in this category constitute an imminent threat to life or limb and should be acted upon within a week. Items in this category should be approved by the maintenance manager. If items require maintenance support the planner/scheduler should make out and log the necessary work orders. The work orders should be signed by the maintenance manager.

NON-CRITICAL SAFETY—Priority 4

Safety items in this category should be accomplished within 3 months. Except for this, the items should be processed as described under Urgent items.

In the event the maintenance manager disagrees with a safety discrepancy, a meeting should be arranged by the manager with the safety engineer for the purpose of resolving differences in opinion. In the event the conflict cannot be resolved at this meeting, it should be resolved at the general manager's or president's level. The safety engineer should prepare a monthly list of delinquent safety items for submittal to the general manager with a copy sent to the maintenance manager.

The safety engineer should be responsible for following-up on all safety items until they are accomplished, or, in the case of contested items, until the issue is resolved by the general manager.

The planner/scheduler should be responsible for reporting the status of safety items to the safety engineer on a weekly basis. He should also be responsible for following-up on safety items requiring maintenance support to insure compliance with schedule requirements of the priority assigned.

16

Scheduling

GENERAL

Scheduling is one of the most important tools used by management for insuring high labor productivity and the orderly accomplishment of tasks.

Scheduling arranges to have the planned elements available *prior* to the job start time.

Typical of the elements required to perform a job are:

- Labor
- Material
- Tools
- Equipment
- Equipment outage
- Safety lockouts
- Instructions

This section describes a recommended scheduling system. All of the required scheduling systems are described. Each individual organization may want to modify, add to or delete from these recommendations. Some person should be assigned to handle scheduling. This may be a foreman, a planner/scheduler or a full time scheduler depending on manpower availability. It is quite common to have at least one person serving as a combination planner and scheduler.

SCHEDULING FUNCTIONS OF THE PLANNER/SCHEDULER (PS)

The Planner/Scheduler has the following scheduling responsibilities:

- Allocating manpower
- Managing capacity and workload fluctuations
- Estimating
- Scheduling of work
    - PM
    - Trouble calls
    - Major maintenance, construction, mods and rearrangement
    - Routine assignments (i.e., janitorial work)
    - Deferred maintenance
- Managing the manpower pool
- Control of backlog

The planner/scheduler should be one of the most knowledgeable men in maintenance about operations in general, operating problems of the day, outages, emergencies, etc., so that he can effectively function as a coordinator between operations and maintenance.

Supervisors should be able to come to him and get answers to the following questions:

- Where is labor assigned in general? This information comes from the monthly labor assignment sheet (See Figure 3-5).
- Total men missing because of absences, vacations, sick leave, transfers, etc.
- Current emergencies
- Location of all files and records
- Location and availability of drawings and manufacturers instructions

	Hours per day
A. Unit Capacity	
10 men x 8 hours/day	80
Overtime	0
	80
B. Routine Assignments	
1. Fixed	
a. PM	8
b. Other	8
2. Variable-trouble calls	8
	24
C. Unavailable Time	
Absentees, vacations, sick leave, training, union business, etc.	8
D. Nonroutine Assignments	
Time available for special work orders A-B-C = D	48

- Each area foreman should prepare a separate labor assignment sheet for each trade.

Figure 3-5: Labor assignment sheet.

- Impending and current outages.
- Work order status

ALLOCATION OF MANPOWER

The first step in the scheduling process is to prepare a labor assignment sheet for each of the area maintenance units. This is the first part of preparing the short range or annual plan. These sheets, which show how labor is distributed by job type, should be updated monthly, if required. The labor assignment sheet is shown in Figure 3-5. An explanation of this sheet follows:

A–Unit Capacity: This is the number of permanently assigned personnel times 8 hours per day. Pooled personnel should not be included in this figure.

B–Routine Assignments: "Routine Assignments" should be divided into two categories:

Fixed Assignments of Known Magnitude

Preventive Maintenance Inspection Route Work—Foremen should go on the route at least once to measure the amount of time it takes. Once this is done, this daily assignment is of known magnitude. Manpower should be allotted to it and should be left undisturbed by other work.

PMI's (Preventive Maintenance Instructions)—Each week the computer prints the week's schedule of preventive maintenance tasks derived from preventive maintenance instructions. This manpower is known beforehand and should be set aside and left undisturbed.

Other Fixed Assignments—There will be other fixed assignments of known magnitude such as oiling, lamp replacement, and janitorial work, etc. This manpower can be calculated and set aside.

Routine Assignments of Unknown Magnitude—Examples of this category are:

- Minor maintenance
- Operating foremen surfaced preventive maintenance

- Emergencies
- Trouble calls

The amount of manpower required for this type of work can be estimated by averaging past experience shown on the blanket work orders.

C–Unavailable Time: This is the average time spent by the crew on absenteeism, vacations, sick leave, coffee time and union business, etc. This figure can be obtained from computer records, past experience and knowledge of the current situation.

D–Nonroutine Assignments: This item is the amount of labor available for nonroutine assignments, and is equal to A minus B minus C. This category of work comprises practically all of the preplanned workload and is the major scheduling task of the planner/scheduler.

MANAGING THE MANPOWER POOL

There may be several separate area maintenance groups, each headed by a foreman. During any one day the workloads of each of these groups will vary. Variations may be caused by breakdowns, changes in priorities, jobs, inability to obtain outages, and a wide variety of other reasons.

Furthermore, the crew capacity available to perform the work will also vary for the following reasons:

- Absences
- Vacations
- Sick leave
- Transfers to emergency jobs
- Transfers to higher priority jobs
- Inability to shut down equipment for servicing (Prevents use of labor)
- Terminations

- Union business
- Inability to obtain material or equipment (Prevents use of labor)

It is the responsibility of the planner/scheduler to be aware of these variations and to make adjustments to balance load and capacity.

Adjustments can be made by one or more of the following methods:

Insufficient Capacity/Excess Load

- Overtime
- Outside contracting
- Delay, reduce or cancel work
- Inter-unit transfers
- Allocation of manpower pool labor

Excess Capacity/Insufficient Load

- Schedule more backlog work
- Schedule deferred maintenance tasks
- Preschedule selected preventive maintenance tasks. (The planner/scheduler can add workload by scanning next weeks preventive maintenance tasks, particularly quarterlys, semi-annuals, and annuals. It will not hurt the preventive maintenance program to schedule an annual overhaul, for example, one week sooner.)

IT MUST BE ASSUMED THAT THERE WILL BE TIMES WHEN ONE MAINTENANCE UNIT IS OVERLOADED WITH WORK AND ANOTHER IS UNDERLOADED. IF NOTHING IS DONE ABOUT THE SITUATION, WE CAN EXPECT THAT THE UNDERLOADED UNIT WILL WORK AT LESS THAN PEAK EFFICIENCY. FURTHERMORE, THE OVERLOADED UNIT MAY HAVE TO DEFER WORK THAT IS CRITICAL TO PRODUCTION.

CLEARLY EITHER SITUATION IS TO BE AVOIDED. THE SOLUTION TO THE PROBLEM IS TO PROVIDE A MANPOWER LABOR POOL.

The question is, how big should the pool be?

The answer to this question can be obtained by first determining the size of each area maintenance unit. Each area maintenance unit should have enough manpower to take care of the average labor required for routine assignments and unavailable time (items B & C of Figure 3-5).

In addition there should be labor set aside in an amount equal to the *average of the minimums* of the variable nonroutine assignment workload (item D of Figure 3-5).

The method of determining the fixed, permanently assigned portion of each area maintenance unit and its contribution to the manpower pool is shown in Figure 3-6.

The difference between the average and the average minimum nonroutine labor quantity will be the particular area maintenance unit's contribution to the manpower pool. This amount of manpower will routinely remain in its home area maintenance unit and be assigned elsewhere only when the planner/scheduler decides there is a need. In other words this isn't a pool of men in the traditional sense that normally gathers in one place awaiting assignment. Stated differently this arrangement constitutes a preplanned agreement between the maintenance manager, his foreman and the planner scheduler that certain men are permanently assigned to a unit and others are subject to redistribution to other units on a daily basis.

With this method of determining the area maintenance crew size, there should *always* be a sufficient workload for the fixed staff. In addition, there will be a pool large enough to handle the majority of labor fluctuations without borrowing continuously from the fixed area maintenance staff.

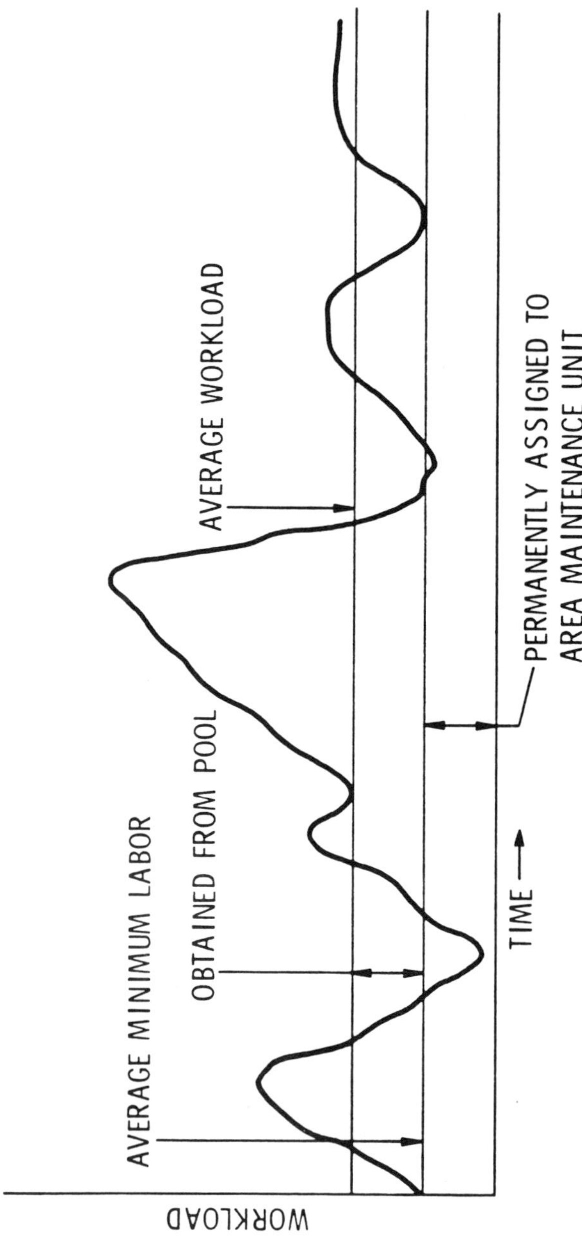

Since there will be times when one maintenance unit is over-loaded and another underloaded, a manpower labor pool should be provided for non-routine labor assignments to absorb the brunt of workload fluctuations.

Figure 3-6: Manpower pool determination.

HOW PREVENTIVE MAINTENANCE INSTRUCTIONS ARE SCHEDULED

The scheduling function for preventive maintenance instructions is performed by the computer as described in the section on preventive maintenance. All preventive maintenance instruction cards should be distributed by the planner/scheduler each Friday afternoon to the appropriate area foremen. The area foremen should have all of the following week in which to accomplish the preventive maintenance instructions. There will be little involvement of the planner/scheduler work other than to coordinate outages if required.

As a result of the load balancing effort of the computer, the manpower required for preventive maintenance instructions should be relatively constant and even. Consequently, a fixed staff should be set aside for this function. In the case of insufficient preventive maintenance instruction workload, men can be assigned to backlog work. In the case of a momentary peak caused by perhaps a large annual overhaul, men can be borrowed from the manpower pool.

HOW TROUBLE CALLS ARE SCHEDULED

As previously stated, the average manpower expended on trouble calls should be determined and set aside in each area maintenance unit.

In areas where the workload is small, the fixed crew can work alternately on preventive maintenance instructions, trouble calls, fixed assignments and daily Inspection Route Sheet Work. The detailed every day scheduling of men to this type work should be handled by the area foremen rather than the planner/scheduler. In this regard the planner/scheduler will be involved only in major workload/capacity fluctuations.

Trouble calls should be received at a "trouble call desk," which should have a separate telephone line set aside just for this purpose. The trouble call desk may be manned by a clerk, who will receive and dispatch trouble calls to the area foreman.

Trouble calls which appear to be over the dollar or hour limit of the appropriate trouble call blanket work order, should be referred by the clerk to the planner/scheduler for input into the work order system.

It should be the responsibility of the foremen to periodically audit the performance and work assignments of the trouble call mechanics to assure that they are:

- Working with reasonable productivity.
- Working on approved types of work.
- Handling an appropriate share of the trouble call workload.

The planner/scheduler should analyze the job type and quantity of trouble calls. The following information should be obtained:

- Highly repetitive trouble call areas or types.
- Average time per trouble call per type.
- Numbers of calls per month.
- Average hours per job type by employee.

All of these statistics can be useful in surfacing trouble spots, poor performance and in scheduling of work.

HOW FIXED ASSIGNMENTS ARE SCHEDULED

The scheduling of fixed assignments is a simple task accomplished as follows:

1. Define task
2. Estimate manpower required
3. Assign estimated manpower to task
4. Periodically audit performance to task including:
    - Task completion
    - Work quality
    - Conformance to estimate

Examples of fixed assigments are:

- Janitor's work
- Daily inspection route preventive maintenance work

Using the daily inspection route preventive maintenance work as an example, the planning and scheduling task would proceed as follows:

1. Define task—This is done by filling out the preventive maintenance inspection route form.
2. Estimate manpower required—This can be done simply by walking thru the task with a competent mechanic working at a reasonable rate to see how long it takes him. In the case of a janitor, the task can be estimated from detailed work standards.
3. Assign estimated manpower to task.
4. Periodically audit performance—The form filled out by the mechanic tells what he was supposed to have performed; the foreman should periodically audit on an after-the-fact basis if he did what he signed off as having completed. At the same time the foreman should check the actual vs. estimated time and the quality of work. In the case of the janitor the foreman can periodically check the janitors performance to task, quality and estimate. He should also take note of customer complaints. Lack of these auditing steps invites poor performance when the maintenance worker discovers that no one ever checks to see whether or not he is doing his job.

The point to be remembered on fixed assignments is that if the 4 steps are not followed you can't tell:

- If the task is being accomplished
- If the quality of work is satisfactory

- If the productivity is adequate
- If the correct amount of labor is assigned

HOW PLANNED JOBS ARE SCHEDULED

The scheduling of planned work orders will be the major scheduling responsibility of the planner/scheduler. It consists principally of major repairs, construction, modification and rearrangement work.

In order to systematize and clarify the scheduling process a planner/scheduler scheduling system should be installed near the planner/scheduler's desk. There are many types of systems that may be used. A very simple system which can be used is shown in Figure 3-7. This system incorporates all of the elements required in a scheduling system. The device is simply a place to visibly file work orders as they proceed through the various scheduling steps. Anything that serves the same purpose will do.

Referring to Figure 3-7, the scheduling process can be shown by describing the operations taking place as the work orders are manually moved through each of the 7 slots in the device.

Step 1—Incoming Work Orders

Upon initiation of the work order, all copies should be placed in this box. The work order at this stage should have all information called for except the planning information.

The planner/scheduler then decides whether to defer the work, in which case it should be filed in a deferred work file, or to process it for accomplishment.

If the planner/scheduler decides to process the work order, he then transfers all copies to the appropriate slot in box 2 depending upon who he decides should do the planning. The choices he has are:

- Maintenance staff (This includes the planner/scheduler, the maintenance manager or other staff personnel).

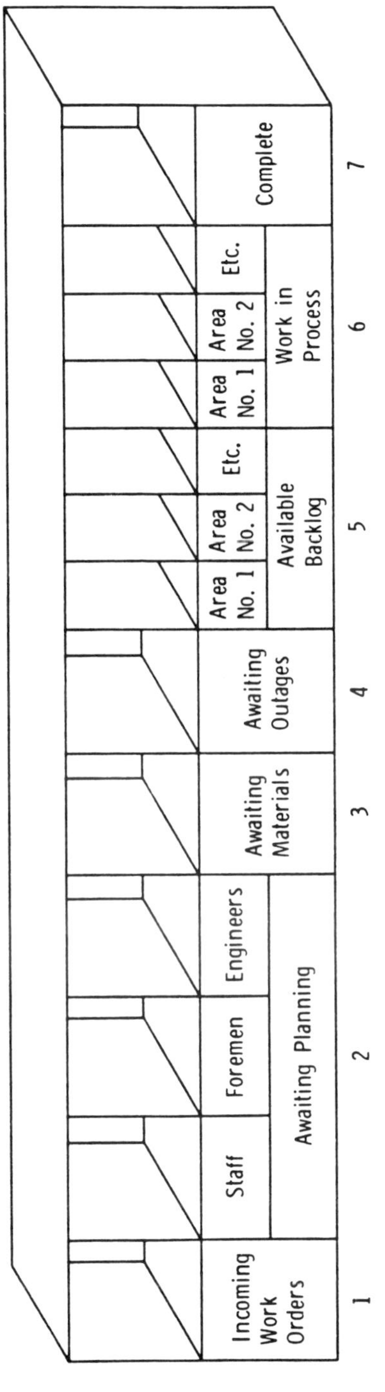

Figure 3-7: Work order scheduling aid.

- Maintenance foreman
- Engineering personnel

After the decision is made as to who should do the planning, the name of the selected planner should be written on the work order.

All copies of the work order should remain in box 2 during the planning stage, except the copy given to the planner.

Step 2—Awaiting Planning

After being assigned to the planning task the planner should accomplish the planning.

Upon completion of the planning task, the planner should enter onto the work order any missing information including the estimate. If necessary, he should rewrite the work requested and job title boxes if the existing descriptions aren't meaningful. The planner may also change the priority upon approval of the planner/scheduler, if warranted.

The planner should attach any auxiliary documents such as the bill of materials, specifications, or drawings to the work order.

Step 3—Awaiting Material and/or Outage

Upon completion of the planning process the planner/scheduler shall:

1. Transfer copies of the work order to box 3.
2. Audit the completeness and accuracy of the work order information including coding and open the work order in the accounting or computer system.
3. Send a copy of the bill of materials and outside purchase requisitions t o the storeroom for performance of the kitting function.

Upon receipt of the bill of materials and the outside purchase requisitions the storeroom should proceed to kit the materials.

The storeroom has the responsibility of completing the kit by the due date established on the bill of materials.

If the storeroom discovers that the due date is going to be missed, they should immediately call the planner/scheduler, so that he can take corrective or alternate action with the help of the job planner.

Step 4—Awaiting Outage

If an equipment outage is required, the planner/scheduler should, upon completion of the material kit, transfer copies of the work order to box 4 and firm up arrangements for the outage called for in the job plan. Ordinarily this will already have been arranged and the work order will be put in slot 5.

Step 5—Available Backlog

Upon completion of the material kit, and obtaining of an equipment outage, if required, copies of the work order should be placed in the appropriate slot of box 5, depending on the area maintenance unit.

These work orders represent available backlog of work.

One of the prime purposes of the planning and scheduling function is to have an adequate "Available Backlog." This is one of the biggest steps forward in achieving high productivity. An empty box 5 means one or both of two things:

1. There are too many people on the staff (not enough work).
2. Some part of the planning and scheduling function is malfunctioning.

Step 6—Work In Process

Once per week, on Friday afternoon, the planner/scheduler should select, for each area maintenance unit, a week's worth of work from the available backlog with careful consideration to the job priority rating.

Upon selection of the workload, the planner/scheduler should send the work order and auxiliary documents to the appropriate foreman. One copy of the work order should be filed by the planner/scheduler in the appropriate "Work in Process" slot for follow up purposes.

The foreman then has the entire next week in which to complete the week's planned workload subject to:

- Priorities
- Due dates
- Prearranged outages

BACKLOG CONTROL

The planner/scheduler should maintain a record of backlog similar to that shown in Figure 3-8.

These graphs will be useful to plan vacations, to facilitate the work planning effort and to make decisions on manpower including the need to hire, fire or alter the numbers in each trade.

In most plants which have their own computer, this information can easily be made and updated as frequently as desired by the computer in tabular or graphic form.

MANAGING CAPACITY AND WORKLOAD FLUCTUATIONS

The workload in any maintenance unit can be subdivided into the following fixed and variable categories:

FIXED

1. Preventive maintenance
2. Fixed assignments

There should be little, if any, backlog of PM or minor trouble call work. Backlog should be primarily deferred, major maintenance and construction work.

The planner/scheduler should maintain a graph of the backlog of work on a weekly basis for each craft. This information will be helpful in scheduling vacations, planning work, and determining allocation of manpower.

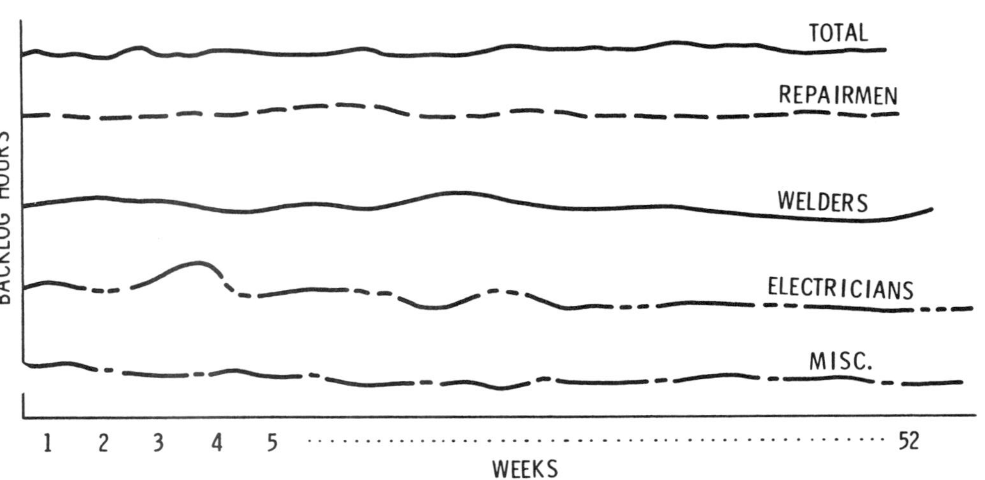

This chart is also used to:

- determine when to hire or layoff personnel
- interchange trades when allowed
- change labor mix
- determine allocation of deferred work or to defer work

Figure 3-8: Backlog of work.

VARIABLE

3. Minor repairs (Trouble Calls)
4. Major repairs
5. Construction, modification, and rearrangement work
6. Absences, vacation and sick leave

The labor capacity available to apply to any of these fixed and variable loads is variable. There are many reasons for this, including:

- Absenteeism, vacations, and sick leave
- Labor transferred to emergency jobs
- Labor transferred to higher priority jobs
- Inability to obtain material or equipment
- Inability to obtain outage

In other words, the labor capacity to meet the needs of any particular load is dynamic and constantly changing. For this reason, it must be constantly monitored and adjusted. The following suggestions are offered to systematize and simplify one of the most difficult problems in maintenance management—the continuous balancing of load and capacity.

(a) First of all each foreman should decide to perform all PM work on schedule. Normally PM work is a fairly constant workload, so a fixed number of workers can be set aside to handle the load. The only variation then will be from vacations, sick leave, and absenteeism. This should be made up as it occurs. There will of course be a temptation to take people away from PM to handle emergencies, however, this should be avoided since it was probably lack of PM that caused the emergency in the first place.

(b) Concerning fixed assignments, decreased labor capacity caused by absences, etc., should be decided upon a case by case basis. Certain fixed assignments must be manned all the time at design strength. An example of this might be a particular oiling route. In the case of an electrician changing light bulbs, on the other hand, the assignment may be dropped for one or two days.

(c) Concerning construction, modification, rearrangement and major maintenance work, workload variations should be taken care of in two steps:

- As previously stated the planner/scheduler should make weekly allotments of the manpower pool to handle workload variations by area and priority.
- At the daily maintenance meeting the planner/scheduler should make adjustments to the manpower pool allocations depending upon the problems of the day.

(d) Finally, the other variable load in area maintenance will be minor repairs, or trouble calls. With this type of work, since the average job is of short duration and since production downtime is frequently involved, it is desirable to have no backlog. This work should be completed *within one day*. This, however, creates a problem. Without a backlog, input must equal output. Since the workload of minor repairs is variable we would have to have a variable work force in order to achieve this. This end may be achieved in the following ways:

- Working overtime
- Borrowing men from the manpower pool

In summary, it is the intention to have the manpower pool absorb the major brunt of load fluctuations because overtime is more expensive and outside help is usually not available at short notice.

DEFERRED MAINTENANCE

From time to time certain tasks originated from any source may be deferred for future accomplishment. Examples of these tasks may be:

(a) Repainting of offices
(b) Cleaning of a cooling tower
(c) Overhaul of a truck engine
(d) Replacement of a corroded pipe line

Possible reasons for deferring this work may be:

(a) Justification is questionable
(b) Heavy Construction Workload
(c) Large backlog of more critical tasks

Most preventive maintenance tasks, for example, may be deferred without immediate impact on operations. However, an indefinite deferment may eventually cause damage to facilities and may result in lost production time. Items of this nature, which are judged to be worthy of future accomplishments, should be filed in a "Deferred Maintenance File" together with a rough estimate on each deferred item.

This log should be used for "fill-in" work when there is excess capacity. However, deferred work should be approved by the maintenance manager prior to accomplishment.

"Deferred" work should *not* be added to the normal maintenance backlog report but should be reported separately.

PROJECT SUPERVISOR CONCEPT

The project supervisor concept is the follow-on to the project planner concept. The project planner is the *one* person responsible for planning a job. The project supervisor is the *one* person responsible for accomplishing the work in accordance with the plan, regardless of the number of maintenance units involved in the job.

A project supervisor should be designated for each construction, modification, rearrangement, or major maintenance job. He should be selected by the maintenance manager. His name should be put on the work order. He should be held responsible for the successful completion of the total job regardless of the number of other maintenance units involved.

The project supervisor should be responsible for coordinating the work of other units and outside contractors with the job schedule; and for delegating responsibility to an acting project supervisor when he is not on the job by virtue of being absent, working at night or weekends or other reasons. This responsibility includes the absolute necessity of adequately briefing the acting supervisor on all elements of the job including unusual and safety problems.

PURPOSES AND USE OF THE LABOR ESTIMATE

Formal estimating should be performed on all non-routine work orders. The main purposes of estimating are:

- To facilitate scheduling. The estimate makes it possible to load each area with a week's worth of work.
- To provide a basis for checking job productivity. The estimated and actual job costs will show up on the completed work order print out. Jobs that were performed significantly above or below the estimate should be investigated for method problems, incorrect estimating or other reasons.
- To provide a basis for work order approval. No one wants to approve a job without knowing what it will cost.
- For backlog control. The summation of each work order estimate provides a measure of the size of the backlog. Backlog control would be impossible without a job estimate.
- To help in controlling low productivity caused by incorrect crew sizing.

Consider how long a garage owner would remain in business, if when a customer asks how much a car repair will cost, he replied, "I don't know, whatever it costs when the job is finished, that's what it will cost you!"

17

Work Accomplishment

JOB FAMILIARIZATION

Job familiarization refers to the foreman becoming thoroughly familiar with a job that he is going to assign to maintenance workers. This function can be divided into three categories:

(a) *Jobs that he has planned himself*—In this case it is not necessary to become familiar with the job since he is the one that conceived it.

(b) *Trouble calls*—In most cases it will be sufficient only for the foreman to be aware of the trouble call. The mechanic will handle these jobs by himself at the discretion of the foreman. In cases of critical breakdowns, the foreman will undoubtedly elect to investigate the problem personally.

(c) *Jobs planned by others*—These jobs, which will be mainly major repairs, construction, modification, and rearrangement jobs, are the principal ones requiring familiarization on the part of the foreman. He must check the completeness of the planning:

— Drawings

— Bill of materials

— Instructions

— Specifications

— Etc.

If the foreman doesn't thoroughly understand the package, he will not be able to properly and completely brief his men and they will have to obtain the missing information themselves at the expense of crew productivity.

INSTRUCTIONS TO CRAFTSMEN

Craftsmen should receive instructions from the foreman prior to proceeding with maintenance work. The instructions will vary by work type, i.e.:

- Fixed assignments
- Trouble calls
- Major construction and maintenance jobs

Fixed Assignments

Examples of fixed assignments might be the performance of a preventive maintenance inspection route or a janitor cleaning offices. Since in both these cases the men will be performing the SAME TASKS EVERY DAY, proper and thorough instruction by the foreman to the craftsman is MORE IMPORTANT THAN FOR ANY OTHER WORK TYPE. Frequently this fact is not recognized because of the low skill level of this work type. The foreman should also:

- Audit performance to his instruction periodically
- Be on the constant lookout for improved methods

The practice of allowing craftsmen alone to break in new workers on a fixed assignment is to be avoided since they are not responsible for worker productivity. Craftsmen may be used to assist in breaking in new employees after the foreman has given the first indoctrination.

Trouble Calls

Foremen should review trouble calls to determine whether or not they want to investigate the job and give instructions to the craftsman, or simply let the craftsman handle the trouble call completely by himself.

Foremen should recognize that trouble call maintenance requires the highest order of maintenance skill, consequently foremen should set aside their most skilled journeymen for this type of work. When this is done, and the foreman is fully satisfied with the competence of his trouble call staff, he can then let them handle trouble calls completely on their own without instructions.

It is understood, however, that foremen will go to the scene of a serious production stoppage or safety incident and personally direct the repair effort.

Major Construction and Maintenance Jobs

The greatest effort in giving instructions to the crew takes place in the case of major construction and major maintenance jobs. The following statement bears repeating!

> IN ORDER FOR A MAINTENANCE TASK TO BE PERFORMED EFFICIENTLY AND ECONOMICALLY, MANPOWER, INSTRUCTIONS, MATERIALS, TOOLS & EQUIPMENT MUST BE AVAILABLE PRIOR TO THE JOB START TIME.

The instructions to the crew should include:

- Technical and schedule aspects
- Discussion on material availability
- Discussion on special tools availability
- Discussion on special equipment availability
- Safety
- Outages and coordination with production

- Portions of work done by others and time phasing thereof
- Schedule commitments
- Lockouts

SAFETY

FOREMEN ARE RESPONSIBLE FOR THE SAFETY OF THEIR MEN!

The foreman's concern over the safety of his crew is necessary to protect them from injury or death and because a serious safety incident may cause:

- Production loss
- Low morale
- Low labor productivity
- Labor relation problems

Foremen should work with the safety department in eliminating safety hazards.

Secondly, foremen should discuss safety when briefing the men on jobs and alert them to unusual hazards of a particular job and what he wants them to do to conduct the work safely.

Safety engineers will, when investigating all accidents, determine if the men have been briefed on unusual safety hazards for the specific job they were working on when the incident occurred.

Foremen should see to it that certain equipment items necessary for the safe conduct of a job are available to the men. Examples include:

- Protective clothing
- Safety glasses
- Hard hats
- Safety shoes

- Breathing apparatus
- Masks
- Safe ladders
- Spark proof equipment in explosive atmospheres
- Etc.

At Kennedy Space Center, which had many hazardous working environments, a green stamp, which was put on the work order folder, was designed. The stamp read, "I have briefed my crew on the safety aspects of this job." (followed by a space for the foreman's signature). The rule on non-compliance to this procedure was a week off the first time the foreman forgot to sign the stamp, and firing the second time. This also applied if the foreman signed the stamp and it was later found that he didn't brief his men. This severe policy was made necessary as a result of several serious accidents.

PLANT CLEANLINESS

Plant cleanliness is an important responsibility for everyone. An unclean plant may cause one or more of the following problems:

Low Morale—Nobody likes to work in an unclean environment

Production Loss—Ore spillage, for example, may result in damage to bearings, belts, skirts, etc., and eventually cause a shutdown

Health Hazard—Uncollected fumes, dust from spillages, etc., may result in medical problems

Safety Hazards—Fuel or oil leaks may result in slipping and falling

Fire Hazards—Uncollected trash, papers, oily rags, or other combustibles may create a fire hazard

Low Productivity—An unclean plant creates an atmosphere of disorderliness and confusion and may lead to low productivity.

For these reasons it is important that each maintenance foreman makes his contribution to keeping the plant neat, clean and orderly and that he invite his men to do the same.

He should do this by developing a plan. The following suggestions are offered:

(a) Shop orderliness—Foremen should outfit themselves with whatever bins, cabinets, tool racks, etc., necessary to insure an orderly shop arrangement.

(b) Workers should be instructed to keep shop neat and clean.

(c) Workers should be instructed to clean up after each job, i.e., return tools and equipment to shop and clean up dirt, trash and debris generated during job.

(d) Foremen should evaluate the problems or damage to equipment caused by uncleanliness, spillages, etc., and coordinate with production on eliminating the problem. If foremen have difficulty in solving this problem, they should alert the Maintenance Manager to the problem. The Manager should then try do resolve the problem with the General Production Superintendent.

FEEDBACK INFORMATION

The foremen are directly responsible for the accuracy and completeness of feedback information concerning maintenance failures.

The subject is covered in detail in Part 2, however the following points bear repeating:

- The primary *source* of feedback information is the maintenance craftsman who observes specifically what was wrong when repairing equipment. He should enter this information on the Work Order.

- The primary checking for accuracy and completeness of feedback entries is the responsibility of the maintenance foreman.

MAINTENANCE ANALYSIS

Maintenance foremen are responsible for analyzing and correcting repetitive failures on equipment that is under their authority.

Foremen should solicit the aid of maintenance craftsmen in helping him with this function by asking them to keep their eyes open in the field and to be sensitive to problems they personally witness.

TIME REPORTING

Foremen should do whatever is necessary to support the accounting system designed by top management. The information collected by this system is essential for maintaining control of the company.

The foreman's contribution to this system is to insure that all labor expended in his group is properly charged and coded.

CHECKING OF COMPLETED WORK

Foremen must check the accuracy, completeness, quality and timeliness of work completed by maintenance craftsmen. The extent of this check varies with the circumstances and depends on the good judgment of the foreman. Figure 3-9 is a guide illustrating accepted practice. There is, of course, no rigid rule for where, when and how often to check work.

It depends on many factors, including:

- Type of work
- Skill of the workers

TYPE OF WORK / TYPE OF CHECK	NO CHECK	DAILY CHECK	SPOT SAMPLE (AUDIT)	CHECK AFTER COMPLETION	CONTINUOUS SUPERVISION
MINOR TROUBLE CALLS	X		X		
MAJOR TROUBLE CALLS		X		X	
EMERGENCIES				X	X
MAJOR REPAIRS		X		X	X
MINOR CONSTRUCTION				X	
MAJOR CONSTRUCTION		X		X	*
PREVENTIVE MAINTENANCE	X		X		

* WHEN NECESSARY

Figure 3-9: Checking completed work.

- Job criticality
- Time available for checking

The important thing to remember is that checks must be made to whatever extent necessary to satisfy the foreman that his group is performing with excellence.

CASE STUDY 3C: The Linecrew

The example which follows is perfect to summarize this section on productivity control, since it illustrates more of the factors affecting productivity than any true incident I have experienced in my lifetime.

The author was called upon to do a study of the linecrew by the president of a utility company. The function of the crews was to install new transmission lines, new electric service to homes, commercial office buildings and industrial plants and to maintain the electrical distribution system.

I started, as was my practice, by examining the files, work orders, touring the facilities and meeting all of the personnel involved, who gave me the expected promise of their full cooperation and support in my effort. I found very little of value in the files and proceeded to design a form that each foreman was asked to fill out every day, for each job, for two months. In preparing the design of the form, a list was made and coded of all the different types of jobs that the linecrews performed. Examples were:

S1 = Install service to home

S2 = Install service to industrial plant

S2 = Install service to commercial building

P1 = Install type A pole

P2 = Install type B pole

T1 = Install type A transformer

T2 = Install type B transformer

The complete list coded seventy-five different types of jobs.

The purpose of the form was to record certain information primarily for the period when the crew arrived at the site and started work on the job. A partial list of the information I wanted was:

- Crew size and trade mix
- Vehicle number and type
- Total man hours for each job
- Material cost.

In addition, I intended to ride with the crews during the two month data gathering period to determine what the crew was doing when at and not at the job site. The following is a description of a typical day spent with the linecrews:

Background

There were only two crew sizes, 6 man or 13 man crews, depending on the job size. The smaller crew typically had the following vehicles:

- Foremen's passenger car
- Material truck
- Winch truck
- Hole digging truck.

The larger crew had a proportionately larger number of vehicles. Working hours were 8:00 a.m. — 4:00 p.m., with a half hour allowed for wash-up at 3:30 p.m. and lunch from 11:30 a.m. to 12:00 noon. The total number of crews was twenty.

Material Acquisition

At the end of each day the foremen made out a list of material needed for the next day. The material trucks were parked side by side each afternoon at quitting time at the supply building loading dock. In the morning all crews arrived at the loading dock; simultaneously the supply room doors opened and the supply clerks started filling the orders and placing the materials on the loading dock. Each crew would then load its material from the supplies placed on the loading dock.

I discovered to my surprise that the supply building was opened 24 hours/day, 7 days/week. "Why doesn't the supply building night crew load the material trucks?" I asked. "They don't know how to do it properly," was the answer. It took anywhere from 10 minutes to an hour to load the trucks. Naturally, no truck was allowed to leave until all material trucks were loaded, even if one crew took only 10 minutes to load its truck.

Coffee Time

At about 9:00 a.m., all crews departed for coffee! There were about 6 coffee shops around town that all men customarily attended. It was a pleasant way to start the day. There was high morale and a great spirit of comradery amongst the crews. There were the usual cliques, of course. For example, crews 1, 3, and 7 were very friendly and had coffee in the same favorite spot each day. This closeness was warming to witness. However, on my first day out with the crews, I thought that it was a shame that crews 1 and 3 had to have coffee at the usual spot, 5 miles *east* of home base, when their first job was 15 miles *west* of base! The job location, whether it be north, east, south or west, apparently had no influence on their basic need to have coffee at the same spot each day.

"Where the Hell Is the Place?"

Next, the crew took out the map sketched by the design engineer. Some of the engineers were good at drawing maps; most were not. "Where the hell is the place?" became the standard question. I made a note to talk to the chief engineer about map making.

After the Job—What's Next?

I will analyze what happens on the job after this paragraph. On the first day I was surprised to learn about the following procedure that was practiced after the first job was complete:

- How long will it take to travel to and complete the next scheduled job?
- If we tackle the next job, can we get back to home base in time to have half hour clean-up time?
- If the answer is yes, then the crew proceeds to the next job.
- If the answer is no, then the crew proceeds back to home base where they wash up and go home.

The result of this standard practice was that crews quit work anywhere between 1:00 p.m. to 4:00 p.m.

The information gathered over the two month period from personal observations and the special forms designed showed that, on the average, the crew's time was spent in thirds as follows:

1/3 Working

1/3 Traveling

1/3 Idle.

The Job

The period the crews were actually working was studied by the following technique:

1. Determine before job start, the estimated time duration of the job.
2. Select random times from a table for the period of job performance, i.e., 12:07, 12:13, 12:14, etc.
3. At the random time points, count the number of workers working as opposed to being idle.
4. The numbers working were then added and divided by the number of observations to obtain the average number working.
5. The productivity was then calculated by dividing the average number of men working by the crew size.

The percentage of time working from this productivity measurement showed that the crews had an average productivity of 30%. But, they were only working a third of the time, therefore, the overall productivity was 1/3 × 30% or 10%!

Analysis—General

My job now was to analyze all of the data collected over the two month period and recommend corrective action. The president was shocked by the 10% productivity and felt certain I was mistaken. It took me about four hours to convince him of the accuracy of my calculations.

Analysis—Job Type Data

Calculations showed that 80% of all work performed by line crews was making service connections.

In watching a service being installed, I concluded it was basically a three man job. One man climbed the pole with his climbing spikes. He worked on installing the transformer and line connection. Another man worked on the connection at the facility. One man was needed on the ground to feed materials to the two men and operate the winch truck to raise the transformer. Based on this analysis, and the fact that service connections amounted to 80% of all work, I recommended that three man crews become the standard size. Larger crews, if needed, would be made by combining two or more three man crews. Furthermore, I recommended the four vehicles presently used by the six man crews be replaced by one bucket truck for all service connection work, thereby saving three vehicles. The three men could easily fit into the bucket truck cab. I also recommended that the foreman become a roving foreman used to spot check several crews instead of having one foreman for each three man crew. The bucket truck could be used not only to lift a man up to the proper working position, but also to raise the transformer. This was a much superior method than using spikes for the men and a winch for the transformer.
The data also revealed that the 13 man crews were occasionally used on service connections which left a minimum of ten men idle.

Furthermore, it was noticed that many jobs were performed by only one man from a 6 or 13 man crew, while the others were idle.

Analysis—Material Acquisition

It was recommended that the night shift stores crew be trained in loading the material trucks and that they load the trucks at night as soon as they were properly trained.

Coffee Time

It was recommended that the practice of stopping at a restaurant be discontinued and, instead, the crew should bring coffee in individual containers or a large single thermos bottle, and the time taken for coffee be periodically audited to insure that only ten minutes be taken in the morning and afternoon instead of the present practice of taking one half to one hour twice per day.

Time to Go Home

It was recommended that the existing outrageous practice of deciding when to leave for home base be discontinued and the crews should leave instead in enough time to allow for half an hour clean-up time.

Maps

It was suggested to the chief engineer that copies of existing sectionalized maps be used to locate jobs and the engineer place a red "x" on the map showing the job location. In addition, of course, he would include the address. The present practice of having engineers draw maps for each job would then be discontinued.

Maintenance

Because of the large influx of people into the city, priority was given to new construction. The result of this was that practically no maintenance work was being done except to handle emergency breakdowns. As a result, the distribution system was in very bad condition and in need of major maintenance repair.

I suggested that the new improvements to productivity should release many men to handle the huge backlog of deferred maintenance work and any layoffs resulting from an increase in productivity be delayed until the backlog was worked off.

Part 4

MAINTENANCE SUPPORT SYSTEMS

18

General

Parts 1, 2 and 3 have presented all that the author considers a unique contribution to maintenance management.

In Part 4 selected maintenance support functions will be discussed. A great deal has been written in these areas and therefore the remarks are confined to matters I have learned to be significant in the functions of engineering, labor relations, training, accounting, computers and maintenance support systems. In the case of maintenance materials and spares, it was found that most maintenance organizations do not implement the large number of systems that are involved to effectively manage this function. Consequently, these systems are reviewed and listed as a refresher and a few helpful hints are provided for management of the material function. This function is the single factor that most commonly is the cause of low maintenance productivity.

19

Maintenance Materials and Spares

GENERAL

It has been the author's experience in making management studies that material is the single biggest maintenance support function contributing to low maintenance productivity. Why this is true is a puzzle, since material management is a well matured discipline.

Whatever the reason, it is believed that efficient management of the material and spares function is important enough to warrant space in this book. Three of the most common errors committed in material management for maintenance are:

- To have no system of any kind.
- To have no kitting system.
- To not recognize the many different material sub-systems and to leave out management for one or more of these systems. (A list of these systems is given below.)

Three elements are required for the successful completion of a maintenance task; Manpower, Instruction, and Materials —"MIM". This chapter deals with the third element — Materials.

It is important for each foreman to be familiar with all of the systems available for purchasing, stocking, expediting, and obtaining materials in order for him to process his maintenance task with maximum efficiency and labor productivity.

An unavailable one dollar part may easily put four maintenance mechanics out of work for hours while the part is identified, located and delivered. The systems described in this section are designed to minimize this loss. These systems are:

Receiving
Kitting
Main stockrooms
Sub-stockrooms
Bench stock—Free issue
Bench stock—Controlled
Residual materials
Surplus equipment
Scrap
Reparable tools, parts and equipment
Expediting
Inventory management
Stock catalog
Spare parts manual
Material acquisition time

The interrelationship between maintenance foreman and material management personnel for each of these sixteen items is outlined in the following paragraphs.

PROVISIONING

Provisioning is defined as the identification, requisitioning, and stocking of parts that are advantageously stored in the warehouse.

The foreman should recognize there is a cost involved in stocking material over and above the purchase cost. This cost is the return on the money spent for stocking material which would be realized if the money were invested. Because of this the foreman should only request stocking of a part when it is apparent that the savings in equipment downtime and labor productivity indicate an advantage to the company.

With this in mind foremen should submit provisioning requests to the warehouse under the following circumstances:

(a) When new equipment is installed, a recommended spare parts list of items to be stocked should be submitted preferably in advance of the start up date.

(b) When high usage on a currently used part indicates that it should be stocked.

When submitting provisioning requests, the foreman should also indicate if the part is to be carried as:

(a) General stock and spares

(b) Insurance spares

(c) Critical dedicated spares

General Stock and Spare Parts are items used generally throughout the plant and are normally of low dollar value and high usage. When provisioning these items foremen should include a recommended minimum-maximum quantity and an estimated annual usage rate to the chief warehouseman. After the item has been placed in stock it should be the responsibility of the chief warehouseman to monitor usage, and recommend and process changes to minimum-maximum figures to maintain proper stock levels and to circumvent stock depletions.

Once a year, the chief warehouseman should list items falling below his recommended minimum annual usage rates and decide on corrective action in cooperation with the maintenance and production managers. Possible actions taken will be:

(a) Delete item from stock

(b) Reduce minimum-maximum levels

(c) Reduce minimum annual usage rate

Insurance Spares

Foremen may decide, in the case of production critical equipment, that spare parts should be stocked regardless of the annual usage rate. For example a set of bearings for a very large motor may be stocked even though the anticipated usage is one set per 4 years.

In this case, the cost of equipment downtime is the overriding factor and takes precedence over inventory expense. These spares are called "insurance spares".

The foreman should designate the quantity to be stocked and the warehouse should reorder on a "one for one" basis. That is, if one is withdrawn, one is ordered to replace it. No minimum annual usage rate control should be applied to this type of stock.

Critical Dedicated Spares

A particular maintenance foreman may have responsibility for an item of production equipment that is so critical he doesn't want to chance having someone else requisition and possibly deplete the stock on a critical spare for this item of equipment.

He can guard against this possibility by designating the spare as a "critical dedicated spare" on his provisioning request.

Dedicated spares may only be issued to the person originating the provisioning action or his alternate. The spare will be coded accordingly in the stock catalog and a note made of the restriction on the parts bin.

Dedicated spares will receive the same type of inventory control as insurance spares.

If a foreman other than the one authorized to withdraw a particular stock item needs the item, he must convince the authorized foreman to sign a stock withdrawal slip.

Responsibility of Chief Warehouseman

The chief warehouseman should be charged with the responsibility of insuring that an "efficient" stock of materials is maintained in the warehouse.

In this case "efficient" means not too high and not too low a level. He shall achieve this by monitoring usage rates.

As a part of this responsibility, he should have the authority to reject any provisioning request particularly of items designated as insurance or dedicated spares, particularly if he feels the use of these categorizations is unwarranted.

It should be emphasized that designations of insurance or dedicated stock should be held to an absolute minimum for the following reasons:

(a) Insurance stock represents a relaxing of inventory level control and too widespread a use may raise stock levels to an uneconomical point.

(b) Dedicated spares represent a restriction on stock usage which can hamper operations of people not authorized to withdraw the part.

If a foreman's provisioning request is rejected by the chief warehouseman he may appeal the decision to the next higher level.

Provisioning Procedure

When a foreman desires to stock a new item he should fill out a provisioning request. See Figure 4-1. This document should include the following items:

(a) Description of part including manufacturer's Part number

(b) Suggested vendor

(c) Recommended minimum-maximum

(d) Stock type designation

(e) Estimated annual usage

The foreman retains one copy and forwards original to the maintenance manager for approval. Upon approval, he forwards original to chief warehouseman who:

(a) Accepts or rejects request

(b) Assigns stock number and location

(c) Fills out stock card and warehouse stock purchase requisition

NO. 0742 PROVISIONING REQUEST DATE__________

PLEASE ADD, DELETE, OR MAKE CHANGES TO STOCK AS INDICATED.

ADD ☐ DELETE ☐ CHANGE SPECIFICATION ☐

DESCRIPTION: (Include Mfg's Serial No.)__________

SUGGESTED VENDOR (S):__________

RECOMMENDED MINIMUM ANNUAL USAGE:__________

RECOMMENDED MIN__________ MAX__________

STOCK TYPE: GENERAL ☐ INSURANCE ☐ DEDICATED ☐

THIS PART IS A SPARE FOR:__________

ORIGINATED BY:__________ REQUIRED DATE:__________

APPROVED BY:__________ APPROVED BY:__________
SUPERINTENDENT WAREHOUSE

Figure 4-1: Provisioning request.

(d) Forwards requisition to purchasing for procurement action

The same form should be used to process recommendations for deleting parts from stock.

Maintenance of Stock Levels

Once an item has been approved for stocking, it should be the responsibility of the warehouse to maintain adequate stock levels.

This should be accomplished by observing stock levels while making entries on the stock cards. When a stock level is too low the warehouse should pull the warehouse stock purchase requisition and send it to purchasing for procurement action. Normally the maximum quantity should be ordered; however, if the warehouse has first hand knowledge of an unusual one-time requirement considerably in excess of the maximum storage quantity, he may up the minimum reorder point by this quantity to circumvent long term stock depletion.

PURCHASING

There are four main items of concern relating to satisfactory purchasing service to the maintenance function. These are:

(a) Adequate communications between operation and maintenance departments and purchasing on the criticality of purchase requisitions (PR's).

(b) Adequate system of reporting on status of PR's.

(c) Adequate statistical information for use in measuring the purchasing activity performance.

(d) A reasonable delivery time on repetitive stock purchases.

The purchasing department is a *SERVICE* organization. The success of a service organization cannot be measured only in terms of its internal operating efficiency, but must also be measured by its impact on the efficiency of the customers it serves. The latter is by far the most important. This fact is sometimes overlooked by purchasing agents attempting to reduce labor costs of a few purchasing people at the expense of decreased efficiency of thousands of people they serve. The purchasing agents goal should be the same as that described for maintenance in Figure 1-2 and Chapter 4.

WAREHOUSING

General

The warehouse function is responsible for the orderly receiving, inspecting, storing and issuing of maintenance and other materials. The systems used in carrying out this responsibility are outlined in the following sections. The systems are designed with that old proverb in mind, "A place for everything and everything in its place". Only by making optimum use of all of these systems can we achieve a neat and orderly looking plant while maintaining the most economical use of labor as far as it is effected by the material function.

Receiving

The warehouse receives parts from the following sources:

(a) Outside purchases made for non-stock materials

(b) Outside purchases made for stock materials

(c) Residual parts returned

(d) Reparable equipment, tools, and parts

(e) Surplus equipment

(f) Scrap

Figure 4-2 illustrates the flow of materials into and out of the receiving area.

Incoming scrap should be unloaded into a special holding area for this purpose in the scrap yard.

Incoming surplus equipment should be unloaded into a special holding area for this purpose.

Outside purchases, reparables (repaired parts) and residuals (surplus materials) should be delivered to the main receiving area in the warehouse which should be separated into marked off areas A through E as shown in Figure 4-3.

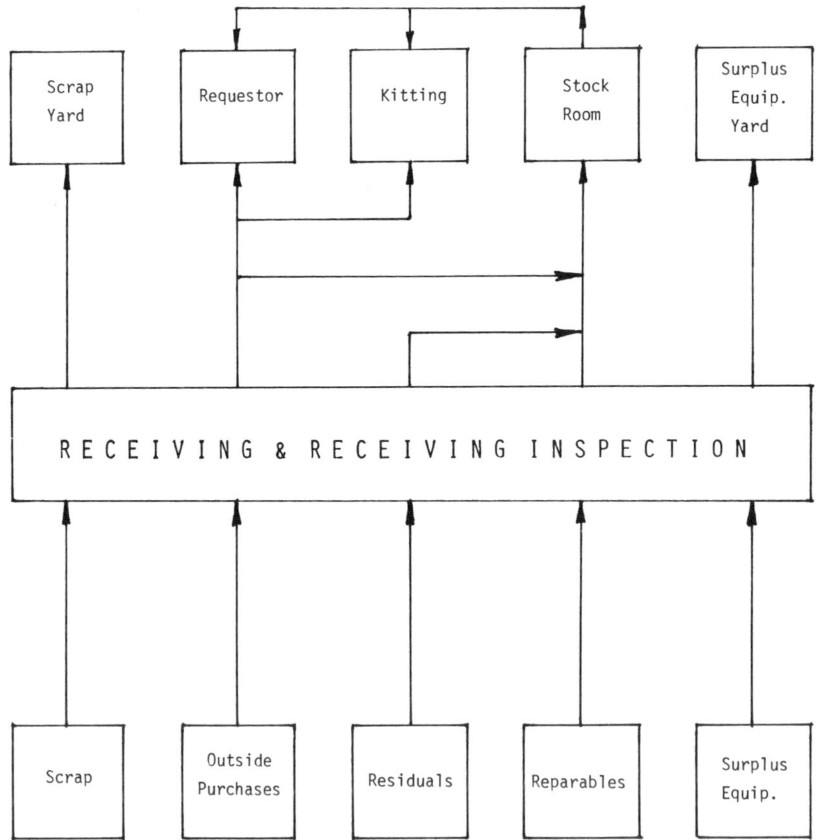

Figure 4-2: Receiving and receiving inspection.

The importance of having these marked off and segregated areas cannot be overemphasized. They create not only a visual impression of efficiency but tend to impart to the warehouse workmen a pride in their area, a sense of order, a visual flow path, and consequentially an actual increase in efficiency.

Kitting

Kitting is defined as the process of obtaining all materials for a specific job in advance of the job start date.

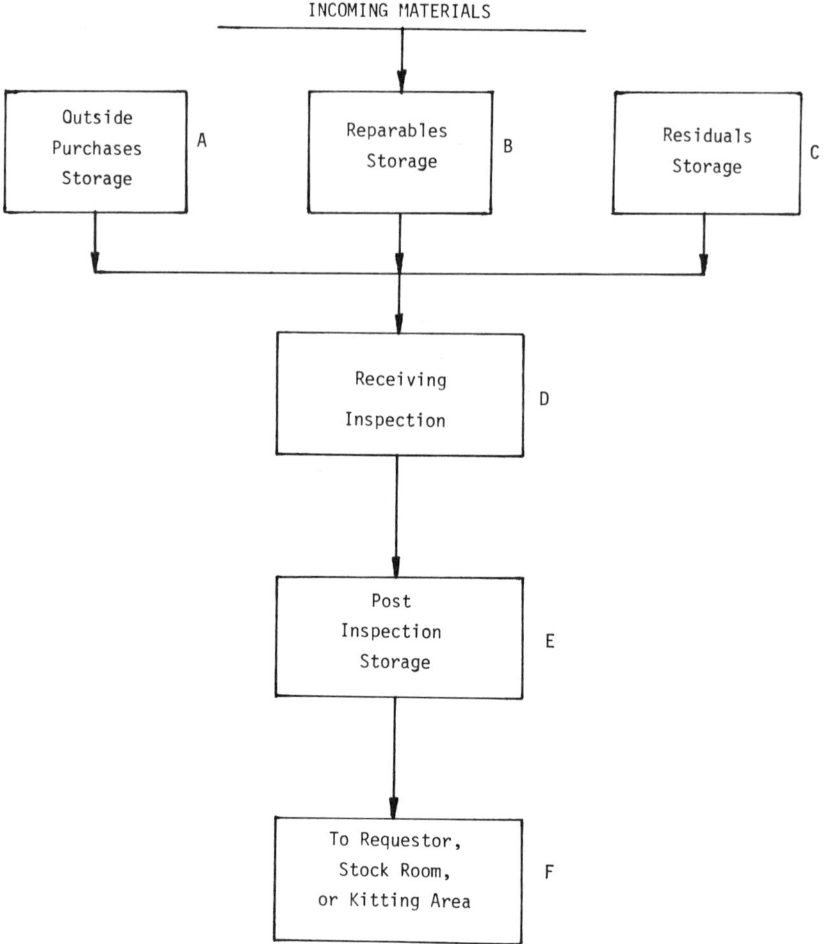

Figure 4-3: Receiving area flow.

A fully "kitted" work order enables the job to proceed without interruption due to material unavailability.

Construction, modification, rearrangement, and major maintenance work (Priority 3 work) can be conveniently kitted in advance due to the fact that this type of work can normally be planned well in advance of its need date. Because of this and the beneficial effect kitting has on labor productivity:

ALL PRIORITY 3 WORK SHOULD BE KITTED IN ADVANCE

Priority 1, 2 & 4 work may be kitted at the discretion of the maintenance foreman.

"Kitting" should be accomplished by utilizing the following procedure:

Foreman's Responsibility

(a) After work order has been prepared and approved, the foreman or engineer should prepare a bill of materials (see Figure 4-4).

BILL OF MATERIALS

DATE__________ WORK ORDER NO.__________

ITEM NO.	QTY.		STOCK NO.	PURCHASE ORDER NO.	DATE RECEIVED

SHEET______ OF______

Figure 4-4: Bill of materials.

(b) Prepare purchase requisition for all outside purchases. (A note should be placed on the purchase requisition to have the purchased item tagged for the work order.)

(c) Send bill of materials to warehouse.

Warehouse Responsibility

(a) Establish manila file folder for each work order which is to be kitted.

(b) Insert copies of bill of materials, purchase requisitions and *purchase orders* in folder.

(c) Write stock requisitions and fill all stock items on bill of materials.

(d) Designate kitting area for work order and mark location on work order kitting folder.

(e) Deliver stock items to designated kitting area.

(f) Receive, check off, and deliver outside purchases to kitting area.

(g) When entire bill of materials is checked off as complete, call foreman and notify him to pick up completed kit.

Foreman's Responsibility

When the work order is complete, the foreman should notify the warehouse who should close out and discard the work order kit file.

Main Stockrooms

Functions normally performed at the stockroom are:

Receiving and receiving inspection

Kitting

Stocking and issuing of materials and parts

Stock part requisitioning

inventory management

Sub-Stockrooms

Under certain circumstances wisdom indicates that advantages can be realized by creation of a sub-stockroom at a remote location. When a sub-stockroom is used the parts stocked may be:

(a) Parts held exclusively in the sub-stockroom

(b) Duplicate stock normally with different minimum-maximum levels

The sub-stockroom may be managed by warehouse personnel in a similar manner to the system used for the main warehouses.

An alternate method of providing a sub-stockroom is as follows:

(a) Locked access to the sub-stockroom will be controlled by the maintenance foreman.

(b) Issues will be made by the maintenance foreman. The requestor will fill out a stock requisition in an identical manner used at the main warehouse.

(c) Stock requisition slips will be forwarded on a daily basis to the chief warehouseman who will post to stock cards located at the main warehouse office.

(d) Reorderinag, restocking, binning and locating of stock in the sub-stockroom will be the responsibility of the warehouse.

Bench Stock—Free Issue

Certain materials are used so frequently and are of such a low cost per unit that it would be uneconomical to purchase or requisition the item from stock one at a time.

An example of this type of item would be nuts, bolts and washers.

The system for handling these items is called *free issue bench stock*. It functions as follows:

(a) The foreman decides what items logically fall within the category of free issue bench stock.

(b) The foreman provides adequate bins and shelves for storage.

(c) The foreman requisitions the parts from stock on a periodic basis and charges parts to the stores account.

(d) The mechanic helps himself to as many bench stock parts as he needs on a free issue basis with no paperwork.

Foreman must be cautious in their decision to identify a part as free issue bench stock and recognize that this action represents a loss in material control and a temptation to personnel who may have use for the part at home. The advantages and disadvantages of this system must be kept in balance and periodic audits conducted of usage rates.

Bench Stock—Controlled

Where the bench system is advantageous but the part is either valuable or most particularly adaptable for private use, the "controlled" bench stock concept may be used. This system is the same as free issue bench stock, except the part is kept locked up either in a cabinet or in a locked and fenced in area. As in the case of free issue bench stock, there is no paperwork transaction, however the person needing the part must obtain the key from an authorized foreman.

Residual Materials

Residual materials are those purchased for a particular task in excess of quantities needed either because of overestimating or because of changed job conditions.

When residual materials are discovered by a foreman he may use one of the following procedures:

(a) Put item in bench stock bin
(b) Scrap item
(c) Return item to stock if it is in units of issue

Surplus Equipment

Surplus Equipment should be defined as assembled, integral units of equipment that are surplus to current operating needs and which are presently not in use. The equipment may have come from one of the following sources:

(a) Change in operating mode
(b) Over order on construction job
(c) Ordered as a spare

The warehouse should be responsible for managing the surplus equipment yard. This yard should be divided into the following areas:

(a) Unresolved area
(b) For sale area
(c) For re-use area

All equipment identified as "surplus" should be tagged with a red tag, identified as "surplus equipment — unresolved", and delivered to the unresolved area of the surplus equipment yard where the equipment should be numbered and logged in by the warehouse attendant.

A disposition of what to do with the unit, i.e. re-use, sell or scrap, should be made by the appropriate area maintenance supervisor.

Surplus equipment to be sold should be tagged with an appropriate yellow tag and located in a separate area.

The warehouse should provide an indexed listing of all surplus equipment available for re-use.

Scrap

The warehouse should be responsible for managing the orderly flow of scrap materials from generation, collection and storage, to disposition.

The collection and storage system should be designed to segregate scrap into logical categories to facilitate sales; for example:

Ferrous
Stainless
Copper
Aluminum
Rubber
Wire
Lumber
Rope
Machinery

Reparable Tools, Parts and Equipment

When a maintenance foreman is faced with the task of putting a broken down item of equipment into operating condition, he has two choices:

(a) Repair the damaged item

(b) Renew the damaged item

When choice (b) is taken he can do three things with the damaged part:

(a) Scrap it

(b) Rebuild it and use it as bench stock

(c) Rebuild it and return it to stock if it is stock item.

Case (c) is the preferred course of action since it will minimize the inventory level and provide for a neater and cleaner plant.

20

Engineering

In many plants the man in charge of maintenance is the plant engineering manager who is also in charge of facilities engineering. Engineering is an important support service to maintenance since it frequently is used to analyze failures particularly when a redesign is necessary and to write preventive maintenance instructions. Both maintenance and facilities engineering are service organizations and therefore the plant engineering manager must please his clients (all the other managers in the plant) if he wants to be successful. For some reason many plant engineering managers recognize that a consulting engineer must please his clients, but never realize that they must too. Some have a tendency to be dictatorial in dealings with their clients.

CASE STUDY 4A: The Newcomer

I had been at work for six months in my new job as plant engineering manager of a large industrial plant with an employment of 50,000 people. I had a staff of 1000 people of which 150 were engineers. It was my first big job. I was only 30 years old, a nervous wreck and starting to crack under the strain. I was the first new manager hired in 20 years. All the other managers were 55 years of age and older. I had a feeling they resented the young "newcomer". To increase my concern, the general superintendent of production, with whom I was friendly, told me that I was in trouble with the old timers. He said all of them wanted a couple of engineers from my central engineering group to be assigned to them permanently to work under their direction so that they didn't have to go through my backlog control to get their jobs done. I explained why I was resisting these requests. I had the younger, less experienced engineers working as "short order" engineers (a term I coined) which meant engineers working as specialists on small, recurring jobs. An example of this work was wiring up new machinery and running new compressed air outlets. We handled thousands of these. As a result of the highly repetitive work these engineers soon memorized the designs and cost estimates so that they could do the work in a fraction of the time that it would take the more experienced engineers I reserved for the larger more complex jobs. I argued that if I acceded to all the requests, there would be nothing left of central engineering and that the decentralized effort with 1-3 people scattered through the plant would require them to know all engineering disciplines, as opposed to the specialist concept I used. This "jack-of-all-trades" concept would mean they would take longer to do a job and the engineering stood a greater chance of being botched. Furthermore, since there would be peaks and valleys to their workload, they might frequently be out of work because of the lack of a large backlog to work to. The superintendent said he was impressed with my argument but warned me that the old timers would probably do me in if I didn't bend a little. I was worried.

Two days later the president of the firm dropped in to see me. He said, "You're in trouble with the old timers." My face turned white. The phone rang. It was someone yelling at me for not having his machine back in service. I stood up and paced back and forth with the phone cord dragging for five minutes as we argued. The president later said he noticed I *always* paced back and forth when I talked on the phone, and it was important for me to project an image of calmness in the face of crisis. I tensed up even more.

He then lectured me for one hour on the need to sell myself to the "old timers." He suggested I visit every manager in the plant once per month with notebook in hand and say, "I would like to know what I can do to improve my service to your department." When I started doing this, they were all stunned! "No other plant engineering manager has ever been interested in my problems!" they all said. I wrote down what they said and did my best to handle their problems. I even gave away some engineers to the managers who were most insistent. In about four months, I was accepted by all and my problems were over.

This was perhaps the most valuable lesson I ever learned and I am grateful to the president of the firm for having taught it to me. The phone incident taught me the need to exhibit a calm exterior at all times, regardless of pressure, a habit which I soon developed.

CASE STUDY 4B: Thirty More?

At a plant where I was Director of Engineering, Operations and Maintenance over 3000 people, the work load was increasing fast. I already had 100 people in the planning and scheduling department and put a request in to management for 30 more. "Thirty more?" they said, "You must be out of your mind!" I explained that the number of jobs handled per month had risen dramatically. "No way will we give you more planners and schedulers," they said.

The scheduling operation was chaotic. Management insisted that *all* jobs be formally planned and scheduled two weeks in advance; the first week being firm and the second tentative. This included time phasing the various trades on projects.

Management wanted us to able to give our customers, instant status on all work. This was a big job in itself. It soon became apparent that every week almost 100% of the schedule was altered because of changes in priorities, absences and other reasons. The entire system seemed like a big waste of time.

I spent the next week tabulating work load data. This showed that 90% of the number of jobs required only 10% of the number of planning and scheduling manhours and 10% of the number of jobs required 90% of the manhours. Furthermore, the majority of the small jobs required less than 8 hours to plan. I saw the solution and made this proposal to management which was accepted:

Small Jobs: On small jobs (90% of the number of jobs), I would set aside 10% of the planners, schedulers and craftsman to handle these jobs. Since the planning averaged less than 8 hours, all of these jobs would be planned in one day with zero backlog. The physical work would be accomplished within two weeks and would not be formally scheduled. Furthermore, no status information would be maintained or given out. The customer would be told simply that his job would be finished in two weeks or less.

Large Jobs: These jobs would be handled with the normal scheduling system complete with formal scheduling, craft phasing and status reporting.

Critical Jobs: These jobs demanded zero schedule slippage and therefore, had to be treated with the highest degree of sophistication. For example, if a control valve needed to be manufactured and delivered by a certain date, the scheduler had to track it through production from the beginning so that he could take action if the production schedule slipped. If necessary to meet the valve delivery milestone date, the scheduler might have to pirate a valve or ask an engineer to redesign it in order to get one on time.

The change in the handling of small jobs allowed us to *decrease* the staffing by 25 instead of having to hire 30 more people! This was the second most important lesson I learned concerning the managing of an engineering (and scheduling) organization. This same principal applies to many other functions.

It can be summarized as follows:

> When a major portion of the number of tasks performed are small, requiring a small percentage of the total staffing, then set aside a dedicated crew to handle the input each day with zero backlog. This makes your clients happy and relieves you of the scheduling and status reporting workload and allows you to concentrate on the few larger jobs.

21

Labor Relations

Americans have been brought up with the concept that there is a natural conflict between the company, represented by management, and labor, represented by the blue collar workers. Each is out to exploit the other.

To a lesser extent, private industry looks upon the government as the enemy, an organization bent upon overtaxing, overregulating and installing restrictive organizations such as EPA, OSHA and the like, so that private industry can hardly stay in business.

Well, look out, private citizens! In countries such as Japan the opposite situation exists! Management and labor work together for each others mutual benefit and the government passes laws and provides aide to private industry to help it be successful. At the end of World War II, the U.S. was the leading shipbuilding, steel making, car producing and electronics nation in the world. This rank has been lost to Japan. Other industries have fallen to Japan and still others are about to.

Maintenance foreman must do their share in inviting cooperation between management and labor to the point it is obvious between both parties that they share a common interest, the *long term* success of the company. Maintaining control over labor relations problem is a function of all supervision even though the overall guidance rests with the director of labor relations. As far as the maintenance foremen are concerned, there are two important points to remember:

(1) *The Labor Relations Triangle:* There should be a mutual feeling of trust, confidence and respect built up between the foreman, the worker and the union steward.

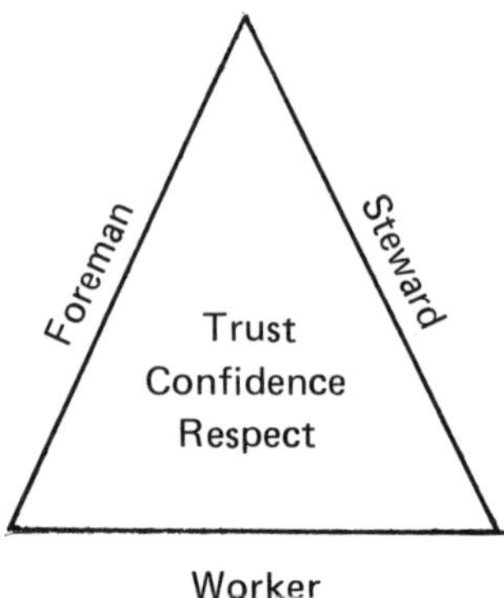

The worker should feel that he can go to his foreman with his problems, and that the foreman will listen to him and give him help when possible. If the foreman doesn't do this, the man will turn to his steward exclusively. Under these circumstances the foreman has lost control of the labor situation and opened Pandora's box.

(2) *Contract Negotiations:* Contract negotiations are not the prime concern of maintenance foreman. Nevertheless, it must be remembered that the unions always have a long range goal (plan) towards which they are actively working. These plans will generally include higher wages, more fringe benefits, greater job security and more jobs. The last two items are frequently achieved by:

- Increased specialization in job titles;
- Decreased flexibility in inter-plant transfers;
- New-hire restrictions.

These inevitably mean reduced productivity and therefore greater employment. The company must have a long range plan to counter these unfavorable trends and to convince the union that their mutual goal is long term success of the company, *and* a pleasant, rewarding working environment for the worker. This is the job of the labor relations department. Nevertheless, they can use help from maintenance foremen who have more direct experience with the operational difficulties caused by restrictive contract clauses and who may have ideas for trade-offs or other suggestions for improvements to the contract. The labor relations department needs this free flow of ideas, suggestions and gripes to be in a strong posture during negotiations. To achieve this level of planning and communication is to be in control of the labor situation.

In addition, the company must recognize that hourly workers are the closest to the operation and its problems. They are therefore in an excellent position to see the problem areas and to recommend improvements. Furthermore, their productivity is a major factor affecting economic success. Their productivity is *heavily* influenced by their morale and attitude toward the company, their boss, their job, working conditions, and motivational programs. These must always be enhanced by a cooperative relationship between management and labor as opposed to the attitude illustrated in Case Study 3B.

In summary, it is critical for us to note that the conflict relationships throughout American businesses that maintain economic and social distance between producers and managers and among firms, unions, government, consumers, owners and financial institutions are virtually absent in Japan. The contrast is startling.

The principal lesson for the United States is that it can no longer tolerate the higher costs of its adversary relationships. In the final analysis, achieving high national productivity may depend more on efficient, harmonious organization than on efficient technology. We are all in this international competition together, and the most socially efficient nations are very likely to be the winners.

22

Training

GENERAL

In Chapter 2 (it was mentioned that the first priority in any maintenance organization is to have a crew of mechanics who can repair anything in the plant that breaks down. This need and priority is obvious. The function that insures this capability, as the plant is modified and updated, is training.

There is an abundance of material available on the subject of training, i.e.: textbooks, plant engineering training manuals, apprenticeship programs, supervisory training, etc. Two unique contributions are offered which are the direct result of observations made in the field of discrepancies in the training program.

At Kennedy Space Center as we were about to start the countdown on the first launch to the moon, I realized we were about to perform certain operational and maintenance tasks that were never performed before. Certain of these tasks were simple whereas others required specialized training. The complex tasks were reduced to written procedures and the men repeated the tasks over and over again until they performed them perfectly. I wanted to make certain this skill was recorded for use by back-up personnel and therefore, developed the technique described in this chapter. It is the author's belief that all plants have a need for this technique for certain specialized tasks.

Another problem noticed in the field while on assignment as a consultant were the large number of plants having complex machines, just installed, without a suitable training program for the mechanic assigned to maintain it.

This gets to be a big problem if the plant has several dozen different type machines without a mechanic who understands how to maintain them. Frequently this problem is the *only* training problem! The company may have an outstanding apprentice, supervisory and other training programs and lack only the training in specific complex equipment. A system for training for this type is described in this chapter.

TRAINING IN SPECIAL SKILLS

Usually there are dozens of special maintenance tasks in a plant that are not only critical to production, but also difficult to perform from a manual dexterity and technical skill aspect. These type of tasks can be taught with the aid of sound movies or video tape.

The technique is similar to that used in the medical profession when they take movies of a surgeon performing some new and difficult operation. Every aspiring surgeon says "If I could only see an expert perform one of these operations, I would feel less hesitant about performing one myself for the first time." The problem may be the particular operation he wants to see is performed rarely and that there are very few experts in the world to perform it. The same situation arises in the maintenance field.

An example will help illustrate how the technique is used. Let us imagine that the special task is the dismantling, overhaul and reinstallation of a gland seal on a turbine. Assume that the seal is very complex with different types of metal, fine finishes, carbon rings, close tolerances, etc. Let us assume that the total task takes 4 hours.

Step 1. Obtain the most qualified mechanic in the plant in this particular task.

Step 2. Obtain a qualified cameraman complete with a Super 8 sound movie camera or video camera. This could be a foreman or mechanic.

Step 3. Take sound movies of the task. Instruct the mechanic to describe what he is doing as he is performing the task, giving special warnings, precautions, and how-to hints as he is going along.

This movie would then be shown to other mechanics who would absorb the expertise of the mechanic making the film.

The wonderful part about this training method is that it is so magnificently inexpensive.

In the case shown above, the only expense is the four hour's time for the cameraman and the film cost. The mechanic's time isn't counted since he is working on an actual job.

It goes without saying that this technique should be limited to specific complex critical tasks where training films are not already available from outside sources. The maintenance manager should decide where to use this technique.

TRAINING IN COMPLEX CRITICAL EQUIPMENT

Let us conjure up a special training problem. Assume that the XYZ Paper Cup Company has only one kind of machine in the entire company, a paper cup making machine. Two types of craft training programs are being considered for the maintenance worker:

(1) A one week training program in maintenance of the paper cup making machine. (A discussion with the manufacturer revealed that this is all the time they need for a complete training course.)

(2) A general correspondence school course in mechanical maintenance coupled with on-the-job training lasting three years.

Clearly program (1) will produce results faster and cheaper in this case.

Almost every plant has cases of equipment that justify specific training in the equipment on hand, in preference to or in conjuction with generalized training.

Specific training of this type should be restricted to equipment items having all of the following characteristics:

- Equipment is critical to production.
- Equipment is complex technically.
- Breakdown repair records show a high failure and downtime rate.

The entire point of this paragraph is to emphasize that in some plants, the heart of the maintenance problem is maintenance know-how on *very few* equipment items and, when this is the case, specific training is indicated rather than generalized training.

The specific training course should be developed by the plant expert in this equipment and the manufacturer and should contain the following:

- A description of the equipment and its controls.
- A discussion of its criticality and downtime cost.
- Preventive maintenance tasks including overhaul instructions.
- Trouble analysis in this format:

  Trouble?

  Probable cause?

  Remedy?

- Operating instructions including how to operate in various component failure mode situations.

ANALYSIS

Operational and Maintenance know-how of equipment is interrelated. A maintenance man cannot be successful without a thorough knowledge of equipment operation. An operating man must have some knowledge of maintenance problems on the equipment he is operating, at least to the extent that operation affects maintenance problems. Operational abuse resulting from lack of knowledge of the machine can cause damage and downtime.

Figure 4-5 illustrates the learning process that a maintenance man must go through in order to have a thorough understanding of the equipment he is responsible for maintaining. Comments on this diagram follow.

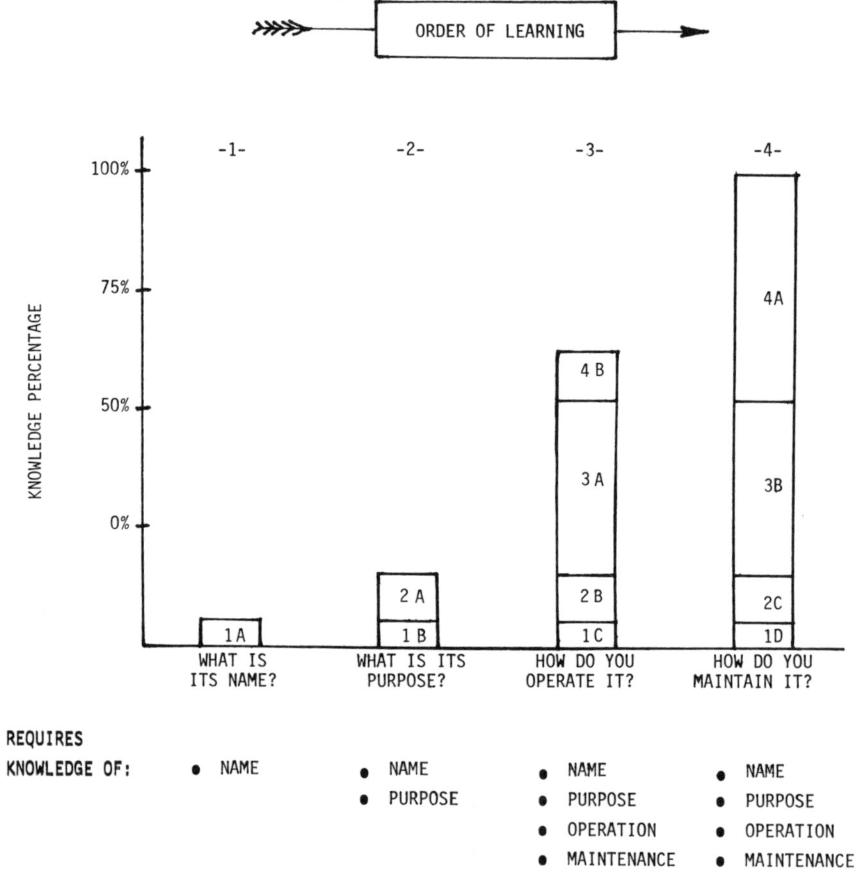

Figure 4-5: Operation and maintenance learning process.

The first step in the learning process is to learn the names of all equipment components. This step (1A), automatically gives an insight to its purpose (1B), How to operate it (1C) and a small idea of how to maintain it (1D).

The next step is to learn the purpose of each component. This step (2A), automatically gives further insight into operation (2B) and maintenance (2C).

The next step is to learn how to operate it. This instruction (3A) does not provide complete operational knowledge since it does not provide for the maintenance impact on operation techniques. However, this training does provide further knowledge of maintenance (3B). The operator must gain the minimum knowledge of maintenance (4B) required to prevent operator abuse.

The final step (4A) is to learn how to maintain the equipment. This completes the maintenance know-how requirement.
It is desirable for maintenance personnel to have 100% knowledge of all four of the discussed items. Operational men should have 100% knowledge of the first three and sufficient knowledge of maintenance to be aware of the operational effect on maintenance.

Many equipment failures have pyramided into excessive downtime primarily as a result of an operators lack of knowledge of the correct operational response when equipment suffers a complete or partial breakdown. It is quite obvious that knowledge is required in the case of an airplane, a ship or a nuclear plant, but the need for it is frequently overlooked in a plant.

23

Accounting and Computer Systems

100% of all the accounting and computer systems I was involved in during my experience as a maintenance and fleet management consultant provided no useful information to the maintenance or fleet manager. Of course my experience only covers a small fraction of the total plants in the United States, however, I cannot help but wonder about the extent of this problem.

The reasons for the lack of usefulness normally were caused by one or more of the following:

(1) The system was designed by the accounting department for its use with no input from the maintenance manager.

(2) The system was designed by a computer expert at the request of the maintenance manager's boss, or at the request of the maintenance manager, with no input from the maintenance manager. In these cases the computer specialist, when asking the maintenance manager what he wants, frequently is told, "I don't know anything about computers, you decide what I can get and should have". The computer specialist then frequently says, "I'll give you everything the data will provide". This inevitably results in

> weekly or monthly computer reports, several feet thick that no one has the time to read.

In many cases there were no systems at all!

The message from this experience is simple:

> Accounting and computer systems designed to provide management information to the maintenance manager *must be designed by him* or by others under his strong technical supervision. *He* is the man who is charged with the responsibility of managing maintenance, and *he* was hired as the top expert on maintenance in the plant and, therefore, only *he* knows better than anyone else, what he needs to properly manage his department.

(My apology to the majority of maintenance and plant Engineering managers who recognize the above advice as being self-evident.)

24

Maintenance Work Standards

GENERAL

"Maintenance Work Standards" is a method of estimating maintenance work using basic industrial engineering principles just as they are applied in production operations. In principle, work is measured with standard time data by breaking down a job into its basic components and applying time standards for each component. The sum of these standard times, plus various allowances for job conditions, represents the time that should be allowed for the job. A comparison of this allowed time with actual time measures the performance or job effectiveness.

"Maintenance Work Standards" or "Universal Maintenance Standards" have the potential of being a very useful tool to the maintenance manager. They also can cause a waste of time and money and a loss in productivity if used improperly, since some people may forget that production operations are highly repetitive while maintenance operations are not.

"THE WORST CASE"

The following example illustrates "The Worst Case" of the use of work standards in maintenance and assumes the standards are applied to all work and every job.

First, the maintenance crew performs the maintenance task and reports the hours spent on the job. This is the "actual time" spent. After the job is complete, the standards applicators are dispatched to the job to calculate how much time the job should have taken according to their set of work standards, which can be a set of books varying in thickness from 1 to 12 inches. This time is the "standard time" spent. The amount of detail used in determining the standard time is very great. For example, if a mechanic takes apart an eight bolt pipe flange, the standards book will show how many minutes or seconds it should take to remove one bolt from the flange under various conditions, i.e.: when the flange is new, in use for five years or when severely rusted. The standards applicator must then add up these hundreds or thousands of time increments to arrive at the total standard time. The standards also provide times for obtaining materials, tools, equipment, travel time, idle time, union business, personal time and many other factors. Standards applicators are frequently industrial engineers with little or no maintenance expertise, and in this case, must first get the maintenance men to show them the job site and explain how the job was performed. This is further complicated by the fact that many of the jobs include opening up a piece of machinery, fixing it, then closing it up. In this case, the standards applicator can't *see* what was actually done, nor the problems encountered inside the machine when it was opened up. Furthermore, the greatest majority of maintenance jobs are of very short duration; because of this and the procedure described above, *it may take more time to estimate the standard time then it actually takes to do the job!* This is particularly serious if the procedure is followed for *all* maintenance work, which is sometimes the case.

The standard time is divided by actual time to give a productivity factor. This information is frequently averaged for each foreman to show the relative ranking of each foreman and to promote a spirit of competitiveness between them so each foreman will strive to obtain the highest productivity factor. This procedure may cost you hundreds of thousands of dollars to implement and millions to operate together with a loss in overall productivity if you don't understand how to properly use and apply work standards.

Remembering that estimating maintenance work is an essential element of an effective maintenance program, and recognizing that engineered maintenance work standards is one of the estimating tools, when and how should we use this tool? The following paragraphs answer these questions.

AUDITING PRIORITIES

When a plant is suffering from low productivity, it is important that the problem be investigated in the priority order described in Chapter 2, Audit Priorities.

The reason for this is that the problem is more likely to be the total absence of one or more of the fundamental management systems, rather than poor estimating practices.

WORK SAMPLING

Work sampling is a very simple tool to use to determine where time is being spent by categories, i.e., travel, material and tool acquisition, idle, working, etc. A work sampling study should be performed at least once per year for each foreman to show if there is an improving or deteriorating trend. For a plant not well organized, this study will show that most of the time spent is non-productive and only a small percentage of time is spent working. Work standards are normally not used to perform the work sampling study. After the study is complete, each of the factors affecting productivity should be carefully analyzed (see Chapter 14) to determine which factor is the biggest contributor to low productivity. This factor should be improved first and then each other sequentially until time spent working is high enough to justify a study of the work method.

ESTIMATING ROUTINE REPETITIVE WORK

Normally, routine repetitive work need be estimated only once, not everytime the job is performed. A perfect example is janitorial work. When a new building is about to be put into service, the estimator makes a list of everything that needs to be cleaned, i.e.: water closets, lavatories, floors, walls, windows, etc. He then uses his janitorial work standards book to estimate the manpower required. There are many of these standards available. Janitorial work represents the best and most useful application of maintenance work standards. After the manpower is selected, then it's up to the foreman to train his men in the work and audit their performance periodically.

Another method of estimating routine work can be explained by using the preventive maintenance daily inspection route sheet work as an example. (The same method can be used for janitorial work). The method proceeds in steps, i.e.:

Step 1

Make a rough estimate using your best judgement.

Step 2

Assign manpower to the task and train them in the work until you are satisfied that they are performing the work properly.

Step 3

Pick the man doing the best job and simply time him to see how long he takes working at a reasonable pace. This time then becomes the estimate and "standard" time.

A final method of estimating may be used on highly repetitive jobs such as installing electrical outlets or compressed air lines. The method starts with recording the actual times on work orders (no estimate is made at all). After a period of time, say six months to a year, the amount of time for all work orders on the highly repetitive task is totalled and then divided by the number of work orders. This is the average time spent and the number to be used in the future for estimating the task. Since these type of jobs are normally of very short duration, and the variables associated with the job high, this method is good enough for scheduling and control purposes. The planner scheduler should do this for all highly repetitive tasks. People who use this method sometimes dictate a small improvement percentage to the standard of perhaps ¼ of 1% over a period of 6 months or a year.

ESTIMATING CONSTRUCTION AND MODIFICATION WORK

The estimating of construction and modification work is a highly developed skill requiring great expertise in many engineering and construction disciplines. There are many reference books, such as *Means Site Work Cost Data 1982*, which are available and give figures for both labor and material. These books are updated periodically.

It is essential that the estimator incorporate, and know how to apply, percentages for escalation and contingency, particularly for large dollar value jobs of long duration. Since construction estimating is such a complex and critical skill which is seldom found in the typical plant engineering and maintenance department, it usually pays to hire outside specialists to do the estimate on the very costly jobs, and at times, use them to check the estimate of the architect/engineering firm or the company design staff. Life cycle costing and value engineering techniques should be used to provide a sound basis for approval of the construction job. Modification tasks employ the same techniques, however, demolition is involved to a much greater extent. In the case of demolition, the estimator must anticipate and study tasks that are "hidden in the walls" lest he leave out a high cost item. Decontamination of hazardous radioactive or toxic wastes are particularly susceptible to underestimating.

ESTIMATING MAJOR MAINTENANCE JOBS

Most major maintenance jobs are basically non-repetitive. Each one is either completely different or at least slightly different than the last job. Parts of the job are known and can be estimated and other parts are unknown and cannot be determined until the machine is opened for inspection as explained in Chapter 15. Major jobs are estimated utilizing a combination of construction estimating aides, past experience on similar or identical jobs and if available, engineered work standards. The most important of these is past experience. There is no easy way to gain this experience other than by hard work, training from experts, the gathering of estimating data, and the passage of time spent working on the estimating task.

ESTIMATING REARRANGEMENT WORK

There are no books on estimating rearrangement work! The man skilled at estimating rearrangement work relies on talents developed in estimating the types of maintenance work previously described and primarily in experience gained in doing this type of estimating for several years and comparing estimated versus actual costs on completed jobs.

ESTIMATING TROUBLE CALLS

The foreman has two options concerning estimating of trouble calls. One is not to estimate them, since the scope of the job is unknown. The other option is to write "guesstimate" on the trouble call slip. The latter option is recommended. It provides the foreman with a rough estimate of whether or not he is giving the mechanic a days worth of work and approximately how long he will be gone on the job. If he estimates the job will take one hour and it actually takes sixteen hours, he is entitled to an explanation. Furthermore, the mechanic should know that someone cares whether or not he spends sixteen hours on a one hour job. If he finds out no one is interested, and if he is not a conscientious worker, you may find him sleeping in some secluded spot. If the mechanic's explanation is reasonable, then you can forget it. If not, you may have to discipline or instruct him. In either case, the mechanic should record the actual hours expended on the trouble call slip so that records may be kept of the total hours spent on this type of work, since the percentage of time spent on breakdowns is a measure of the success or failure of the preventive maintenance program.

SUMMARY

In summary, let me repeat the main points to be remembered when considering the use of engineered maintenance work standards:

(a) When productivity is low, it is normally because of the complete absence, or the inadequate application of the several management functions described in this book, not because of poor estimating techniques. In other words, give the men the tools, equipment, materials, instructions and outages he needs to do the job and arrange to get him to the job expeditiously before you start worrying about how long it takes him to do the job! Study the factors affecting productivity (Chapter 14).

(b) Estimating *is* important! But one must recognize that the estimating system varies with the type of work and that engineered work standards is just one type of an estimating system. Each estimating system must pay for itself in improved productivity. Engineered work standards is one of the more sophisticated systems and, therefore, should be used with knowledge and understanding of its contribution to productivity.

Bibliography

Cooling, W.C., *Low Cost Maintenance Control,* New York: American Management Association (1973).

Corder, A.S., *Maintenance Management Techniques,* New York: McGraw-Hill Inc. (1976)

Godfrey, Robert S., *MEANS Site Work Cost Data 1982,* Massachusetts: Means, Robert Snow Co. Inc.

Gradon, F., *Maintenance Engineering Organization,* New York: John Wiley & Sons Inc. (1973)

Heintzelman, J., *Complete Handbook of Maintenance Management,* New Jersey: Prentice Hall Inc. (1976)

Higgins, L.R. and Morrow, L.C., *Maintenance Engineering Handbook,* 3rd Edition, New York: McGraw-Hill Inc. (1977)

Hildebrand, James K., *Maintenance Turns to the Computer,* Massachusetts: CBI Publishing Co. Inc. (1972)

Husband, T.M., *Maintenance Management and Terotechnology,* England: Saxon House (1976)

Kelly, Anthony and Harris, M.J., *Management of Industrial Maintenance,* Massachusetts: Butterworth (1978)

Lewis, Bernard T., *Developing Maintenance Time Standards,* New York: Farnsworth Publishing Co. Inc. (1967)

Lewis, Bernard T., *Management Handbook for Plant Engineers,* New York: McGraw-Hill Inc. (1977)

Lewis, Bernard T. and Tow, L.M., *Readings in Maintenance Management,* Massachusetts: CBI Publishing co. Inc. (1973)

Mann, Lawrence, *Maintenance Management,* Massachsetts: Lexington Books (1976, 1983)

Morrow, L.C., *Maintenance Engineering Handbook,* New York: McGraw-Hill Inc. (1966)

Newborough, E.T., *Effective Maintenance Management,* New York: McGraw-Hill Inc. (1967)

Ouchi, William, *Theory Z: How American Business Can Meet the Japanese Challenge,* New York: Avon Books (1982)

Patton, Joseph D., Jr., *Maintainability and Maintenance Management,* North Carolina: Instrument Society of America (1981)